FIELD GUIDE
TO THE BIRDS OF THE
EASTERN HIMALAYAS

FIELD GUIDE
TO THE BIRDS OF THE
EASTERN HIMALAYAS

SÁLIM ALI

With 37 Colour Plates
Illustrating 366 Species

DELHI
OXFORD UNIVERSITY PRESS
LONDON   NEW YORK

Oxford University Press

OXFORD LONDON GLASGOW
NEW YORK TORONTO MELBOURNE WELLINGTON
IBADAN NAIROBI DAR ES SALAAM CAPE TOWN
KUALA LUMPUR SINGAPORE JAKARTA HONG KONG TOKYO
DELHI BOMBAY CALCUTTA MADRAS KARACHI

First published 1977
Second impression 1978

Printed in India by
Aroon Purie at Thomson Press (India) Limited, Faridabad
and published by R. Dayal, Oxford University Press
2/11 Ansari Road, New Delhi 2.

PREFACE

The idea of an illustrated book on the birds of Bhutan, on the lines of my *Birds of Sikkim*, originated with His Majesty the late Druk Gyalpo Jigme Dorji Wangchuk, an ardent nature-lover and sportsman. He not only provided me and my colleagues and assistants with all the necessary facilities for travelling and camping in his supremely fascinating but logistically difficult kingdom for several months at a time on three successive collecting expeditions, but also made a substantial financial contribution towards the cost of publication. It is tragic that he should have died while still so young, and before he could see the book he had so enthusiastically sponsored.

On deeper consideration it seemed somewhat redundant to put out a separate book on the birds of Bhutan when one on the contiguous and zoogeographically similar territory of Sikkim was already available.[1] Also when there existed a fuller work on the birds of the Indian subcontinent which covered the entire Himalayan region including Sikkim and Bhutan as well as Arunachal Pradesh farther eastward. In the wider interests, therefore, the Druk Gyalpo readily agreed that it would be more purposeful to produce a handy-sized illustrated Field Guide rather than a conventional book, and that this should cover not only his own kingdom but the entire physiographical unit—the Eastern Himalayas—of which it formed part. I am indebted to the Chogyal of Sikkim and his government for kindly lending for reproduction in this Guide, sixteen of the original plates painted for *Birds of Sikkim*. This has meant an appreciable saving in artists' fees.

Lastly, I must record my grateful appreciation of the friendly cooperation and unfailing courtesies received, collectively and individually, from all levels of the Border Roads Organization in Bhutan. But for their invaluable help and consideration the logistic, commissariat and camping problems in the interior of

---

[1] Since out of print.

this rugged and undeveloped country could well have been insuperable, except with vastly more elaborate and expensive bandobast.

A comprehensive scientific report of the complete orthological survey of Bhutan is under preparation jointly by Dr S. Dillon Ripley, Dr Biswamoy Biswas and myself. This will include the results of the series of recent expeditions undertaken by all the three, jointly and severally, and will be based on our combined collections and field notes.

CONTENTS

NON-PASSERINE

PASSERINE

COLOUR PLATES

INTRODUCTION

'Eastern Himalayas' as considered in this Field Guide is a stretch of extremely rugged mountain country along the northern border of India. The area lies roughly between latitudes 26°30′ and 28° N., and longitudes 87° and 97°30′ E. It is some 1000 km long and varies in width from *c.* 150 to 200 km, including within its boundaries the Kingdom of Bhutan and the Indian States of Sikkim and Arunachal Pradesh—the last forming the extreme north-eastern corner of the Indian Union. The duars of Bengal and Assam, stretching contiguously along the base of the foothills, have been included for the sake of completeness since many of the higher-elevation birds descend to them to spend the winter months. The western political boundary of this area in Sikkim is marked by the Singalila Ridge, or Spur, which rises abruptly from the terai plains and runs about 100 km south to north, separating the Darjeeling district of W. Bengal together with Sikkim State from the Kingdom of Nepal, and culminating at its northern extremity in some of the loftiest mountains in the world, the Kanchenjunga massif, which besides Kanchenjunga itself (8586 m), includes many other peaks of well over 6000 m. However, from biological considerations, and thus for the purpose of this Field Guide, it seems more appropriate to regard, as has been done here, the Arun-Kosi Valley of eastern Nepal, *c.* 100 km west of Singalila, as the natural boundary of the area since it is hereabouts that definite diversity in the character of the avifauna from the rest of Nepal and the western Himalayas becomes sensibly discernible. The geographical position of this mountain belt at the head of the Bay of Bengal—its nearness to the sea and inflow of the moisture-laden SW. monsoon winds till they strike the mountain barrier to condense and cause heavy precipitation—makes it the most humid tract of the entire Himalayan chain. Its lower latitude and relatively warmer climate have conduced to a higher timber-line, higher alpine zone and higher snow-line than in the western Himalayas. Moist steamy tropical valleys occur in the foothills flanked by densely forested slopes seemingly side by side with stupendous snow-capped mountain ranges. The abrupt juxtaposition of so many different biotopes or life zones—ranging from almost plains level to over 6000 m, and from tropical heat to arctic cold—all telescoped within a straight-line distance of hardly more than 80 km, has given to the eastern Himalayas a flora and fauna which for richness and variety is perhaps unequalled in the world. Sequestered in the rain-shadow, moreover, lie dry, practically rainless valleys which add to the ecological complexities of the jumbled habitats

and make the area as a whole particularly rich in flowering plants, butterflies and birds.

Broadly, three main altitudinal zones of vegetation are recognizable,
each the product of local factors—configuration, altitude, rainfall,
temperature and humidity—and diffusing into one another at the seams:

The TROPICAL ZONE. The northern fringe of the Bengal plains
extending northward to the foothills known as the Western and Assam
Duars (Jalpaiguri, Buxa, Goalpara, etc.), is formed of the detritus brought
down by a number of torrential streams, e.g. Tista, Jaldakha, Torsa and
others. It is a humid well-drained tract extremely rich in flowering plants
and ideal for tea cultivation. This zone, which rises roughly to an altitude
of about 1200 m, contains deep-sided valleys and gorges with well-drained
flanking slopes clothed in dense evergreen jungle. It is characterized by
gigantic trees with buttressed boles, with huge lianas and climbers draping
their tops, intertwining in the foliage canopy and festooning from tree to
tree. Orchids of numerous species abound, and wild bananas, straight-
stemmed screw pine (*Pandanus*), nettles and giant bamboos (*Dendro-
calamus hamiltonii*) provide the other features characteristic of this type
of humid foothills forest. Above this belt, and up to an elevation of 2000 m
or so, the vegetation assumes an intermediate sub-tropical aspect, passing
upward into

the MOIST TEMPERATE ZONE. This zone consists of dense, tall
evergreen forest with moss- and lichen-covered oaks and rhododendrons
predominating, and a large variety of other broadleaved species such as
horse-chestnut, laurels, máple, alder, etc. Near its upper limit, i.e. between
c. 2800 and 3600 m, conifers like silver fir (*Abies*), spruce (*Picea*), larch
(*Larix*), hemlock (*Tsuga*), and juniper (*Juniperus*) occur with which are
mixed deciduous trees such as magnolias and poplars; the undergrowth
is made up largely of 'maling' bamboo (*Arundinaria*), dwarf rhododendrons and other evergreen shrubs, and there is a luxuriance of mosses,
ferns, and epiphytes.

The ALPINE ZONE. Several species of rhododendron and juniper of
the upper Temperate Zone extend into the Alpine Zone which stretches
between *c.* 3650 and 5500 m altitude—the limit of tree growth. But the
forest here is composed chiefly of small stunted trees and large shrubs
interspersed with patches of fir and pine. The stunted forest contains a
baffling array of rhododendron species which impart to it an evergreen
character and present a spectacle of unforgettable colour and splendour
when they simultaneously burst into flower. Plant life in this zone is
dependent entirely on snowfall and melting snow, and gets progressively
dwarfed and scanty towards the upper limits. Patches clear of snow
become carpeted with a profusion of herbaceous flowers of many species,

the commonest being the primulas, anemones, *Corydalis*, and *Fritillaria* which add so much to the glory of the Himalayan uplands in spring. The vegetation otherwise is restricted largely to rhododendron, juniper, and cotoneaster scrub. Beyond *c.* 5000 m plant growth virtually ceases, the heights above being covered by rock and perpetual snow. Depending upon aspect, exposure, and other locality factors, the upper 1000 metres or so of the Alpine Zone are often arid and desolate, conforming more closely with the steppe country of the adjoining Tibetan Plateau. Indeed steppe conditions may occur locally at much lower elevations, as in secluded rain-shaded valleys here and there throughout the region, with the accompanying peculiarities in their animal and plant life.

The 536 birds described in the Field Guide, 366 of which are illustrated in colour, make up a fairly comprehensive list for the area. However, a few mostly low-elevation forms such as egrets—also some uncommon Birds of Prey—occurring in the duars and adjacent plains which *may* turn up sporadically in the cultivated upland valleys are omitted. 'Status' and 'Habitat' refer specifically but not exclusively to the Eastern Himalayas. For details of the overall range within the subcontinent, and extralimitally, reference should be made to *Handbook of the Birds of India and Pakistan*.

On the plates, strokes point to the features which should help identification. This useful device, first introduced by Roger Tory Peterson in his world-famous Field Guides, has proved its effectiveness and now become more or less universal in books of this type. On the pages facing the plates the letters **R** and **M** against the names indicate respectively whether the bird is a Resident or extralimital Migrant with specific reference to the area described above.

Calls of birds, and local names, have been accented to aid phonetics. Unless the vowel sounds are correctly pronounced the word becomes unintelligible and serves no useful purpose. The simple diacritical marks used here are as follows:

> ă (short) pronounced like u in cut
> ā (long) pronounced like a in father
> ē (long) pronounced like ea in great
> ū (short) pronounced as in bull

With the exception of the vowel u, a crescent (˘) above a letter means that it is short; a dash (−) above it, long. In the case of u a dash above signifies that it is pronounced short but as in bull and not as in cut. A double o is used to distinguish from it the longer sound oo as in boot.

Approximate sizes of common and well known birds have been found helpful as standards for visual and mental comparison in identification, and are repeated in this Guide. Thus

Sparrow	15 cm (6 in.)	Crow	42 cm (17 in.)
Quail	18–20 cm (7–8 in.)	Kite	60 cm (24 in.)
Bulbul	20 cm (8 in.)	Duck	60 cm (24 in.)
Myna	23 cm (9 in.)	Village hen	45–70 cm (18–28 in.)
Pigeon	32 cm (13 in.)	Vulture	90 cm (36 in.)
Partridge	32 cm (13 in.)		

These measures are merely for convenience and have no scientific import- ance; they refer to the length from tip of bill to tip of tail of a freshly killed bird laid on its back, unstretched; and in the text, measurements from 25 to 50 cm are rounded off to the nearest 5 and thereafter to the nearest 10. By themselves the measurements may sometimes be even misleading since they take no account of features like proportionately long or short bill or tail, or bulk of body. Thus mere cm or in. may quaintly suggest a bird like the Paradise Flycatcher to be the equal of a kite in size! By and large, however, if used with caution such measurements do provide helpful pointers for field identification. Plus (+) and minus (−) signs indicate whether the bird is larger or smaller than the standard, and the sign ± that it is of more or less equal size.

For ease in referring to the *Handbook of the Birds of India and Pakistan* for fuller information than given in this Field Guide, the volumes in which to look for it are indicated below:

Species Nos.	Volume	Species Nos.	Volume
1–32	1	212–270	6
33–61	2	271–346	7
62–111	3	347–414	8
112–168	4	415–484	9
169–211	5	485–536	10

Terms used in the description of a bird's plumage and parts

Topography of a bird

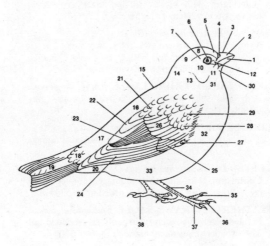

1. Mandible	20. Under tail-coverts
2. Maxilla (or Upper mandible)	21. Scapulars
3. Culmen	22. Tertials
4. Nostril	23. Secondaries ⎫
5. Forehead	24. Primaries ⎬ (Remiges)
6. Iris	25. Primary coverts
7. Supercilium	26. Greater coverts
8. Crown	27. Bastard Wing (alula)
9. Nape	28. Median coverts
10. Ear-coverts	29. Lesser coverts
11. Malar region (malar stripe, moustache)	30. Chin
12. Lores	31. Throat
13. Side of neck	32. Breast
14. Hindneck	33. Belly (abdomen)
15. Mantle	34. Tarsus
16. Back	35. Inner toe
17. Rump	36. Middle toe
18. Upper tail-coverts	37. Outer toe
19. Tail (rectrices)	38. Hind toe (hallux)

Plate 1, artist J. P. Irani
EAGLES, HAWKS, BUZZARD, TUFTED DUCK, VULTURE

0 6 12 18 24 30 cm

JPIrani '72

CORMORANTS: Phalacrocoracidae

1 **LARGE CORMORANT** *Phalacrocorax carbo* (Linnaeus)
Size Duck + ; length 80 cm (32 in.).
Field Characters A glistening black fish-eating water bird. Superficially rather duck-like but with narrow hook-tipped bill and longish stiff tail. Large size, hoary crested head and neck ('pepper-and-salt'), white throat with a bright yellow gular pouch, and white oval patches on thighs diagnostic in flight and at rest.

Non-breeding plumage almost wholly black: thigh patches absent; throat and gular pouch less conspicuous. Young has much white on underside.

Small parties and flocks swimming low in water and diving for fish, or resting on midstream rocks in upright position, wings spread out to dry.
Status Resident, and summer-winter altitudinal migrant. Affects rivers, streams and tarns up to 2000 m.

DUCKS: Anatidae

2 **TUFTED DUCK** *Aythya fuligula* (Linnaeus)　　　　　　p. xvi
Size Duck − ; length 45 cm (18 in.).
Field Characters Boldly contrasting black and white plumage, limp occipital tuft, white wing-mirror (speculum) and bright yellow eye identify the drake. Brown replaces the black portions in the duck. In flight, bi-coloured plumage pattern and broad white trailing edge of wings diagnostic.
Status, Habitat, etc. Transitory. Small numbers seen on high-elevation lakes and tarns while on migration. Recorded up to 5300 m. Food: water weeds, crustaceans, etc. procured by deep diving.

3 **GOOSANDER or MERGANSER** *Mergus merganser* Linnaeus　p. 17
Size Duck + ; length 70 cm (28 in.).
Field Characters Male. In flight general effect black and white, with dark green indistinctly crested head and neck, thin blood-red bill and legs. Female similar but head cinnamon-brown with glistening white throat; back greyish brown; conspicuous black-and-white wings. Identified in flight by elongated grey body, rufous head, white throat,

broad whitish wing-patches, slender neck and thin blood-red bill and legs.

Status, Habitat, etc. May breed in the northern Tibetan facies; chiefly winter visitor. Small parties on clear, rapid streams, more commonly in the foothills but also the upper reaches, locally to at least 3000 m. Food: mainly fish caught by diving. Parties often hunt in coordination. Flight rapid, usually close above the water surface, almost clipping the turbulent ripples.

HAWKS, VULTURES, etc.: Accipitridae

4 BLACKWINGED KITE *Elanus caeruleus* (Desfontaines)

Size House Crow − ; length 35 cm (14 in.).

Field Characters A small grey-and-white hawk with a black line through the eyes (blood-red) and prominent black patches on wing-shoulders. At rest, closed wings extend beyond short (almost square-cut) white tail. This constantly raised and lowered and opened and shut while the bird is perched. In flight resembles Pale Harrier (which also has black primaries), but smaller size, black shoulders and pure white tail diagnostic. Sexes alike.

Status, Habitat, etc. Resident. Duars and foothills up to 1600 m: savannah country, open scrub jungle and grassland interspersed with cultivation. Flight sluggish with deliberate wing beats, alternated with bouts of sailing. Hovers like kestrel to locate ground prey and makes parachute-like descent in steps to seize it in a final rush. Food: large insects, lizards, field-mice, etc.

5 BLYTH'S BAZA or LIZARD HAWK *Aviceda jerdoni* (Blyth)

Size Jungle Crow; length 50 cm (20 in.).

Field Characters A brown hawk with rufous head and erect black occipital crest (depressed in flight). A black mesial stripe down rufous-and-white throat. Underparts conspicuously barred rufous-brown and white. Tail brown with three dark bands, the terminal one broadest and darkest. On a casual sighting confusable with Crested Goshawk (10) at rest; in flight shape and wing action reminiscent of a small buzzard. Female paler above and below.

Status, Habitat, etc. Resident. Confined to evergreen biotope in the foothills between 350 and 1800 m, with some seasonal altitudinal movement. Pairs or family parties of 3 to 5. Rather crepuscular and rarely seen except when soaring in circles, often at treetop height. Call: mewing screams, sometimes confusingly like Palm Squirrel's calls. Pounces on prey from ambush in a leafy tree, carrying it off in the bill. Food: lizards, frogs, large insects, etc.

6 BLACKCRESTED BAZA or LIZARD HAWK
Aviceda leuphotes (Dumont)

Size Pigeon $\pm$; length 35 cm (14 in.).

Field Characters A handsome, crested black-and-white hawk, easily identified by black throat and upper breast with a broad white gorget and a black band below it, boldly chestnut-barred lower breast and flanks, and black abdomen and vent. Sexes alike.

In overhead flight paler underside of tail contrasting with black abdomen and vent, chestnut-barred underparts, and white breast diagnostic.

Status, Habitat, etc. Status uncertain. Breeds up to 1200 m (April-June) possibly migrating south to Kerala and Sri Lanka (Ceylon) in winter to augment the local population. Affects the neighbourhood of streams and clearings in evergreen foothills forest. Somewhat crepuscular. Solos or small parties, often seen soaring in circles above tree tops. Flight sluggish: crow-like flapping with short bouts of sailing. Call: a shrill squeal recalling Pariah Kite. Food: lizards, frogs, large insects— the last sometimes taken by aerial sortie.

7 CRESTED HONEY BUZZARD *Pernis ptilorhynchus* (Temminck)

Size Pariah Kite $\pm$; length 70 cm (28 in.).

Field Characters A variably plumaged raptor, normally greyish brown with darker grey head and short nuchal crest. Underparts pale brown narrowly cross-barred with white. Tail greyish brown with three black bands separated by paler ones. In overhead flight resembles Short-toed Eagle of the plains (also with barred underwing pattern), but distinguishable by slenderer head and neck. Sexes alike.

Status, Habitat, etc. Resident, subject to local movements. Duars and foothills to at least 1200 m: forest glades and groves of large trees in deciduous and semi-evergreen biotope. Usually single, soaring aloft in circles or perched on a tree top. Flies with steady wing beats and occasional bouts of sailing. Food: honey and larvae of wasps and bees, possibly also wax; other large insects and small animals. Call: a single high-pitched screaming whistle *wheeeew*.

The migratory Siberian subspecies (*orientalis*) may also occur i winter. It is distinguished by the black subterminal and median band on tail being much narrower than the paler bands, as against the opposit condition in the resident form *ruficollis*.

8 BLACKEARED KITE *Milvus lineatus* (Gray)

Size Pariah Kite ± ; length 70 cm (28 in.).

Field Characters A large dark fulvous-brown raptor, unmistakable by its deeply forked tail particularly conspicuous in flight. When overhead, a white patch on underwing (coverts), as in Buzzard, likewise conspicuous. Sexes alike.

Status, Habitat, etc. Resident and Migratory. Resident population breeds (March to May) between 1500 and at least 4500 m (observed at 5300 m in May). Migration in March and September. Affects the neighbourhood of upland villages and nomadic encampments, but is less of a scavenger than Pariah Kite. Often seen sailing buoyantly round hill contours and above populated valleys, but is less given to soaring aloft on thermals like vultures. Food: offal, carrion, small or sickly animals, e.g. lizards, birds, rodents, and insects. Largely omnivorous. Call: a shrill musical squealing whistle *ewe-wir-r-r-r* and variants, uttered at rest and on wing.

9 GOSHAWK *Accipiter gentilis* (Linnaeus)   p. xvi

Size Pariah Kite ± ; length ♂ 50 cm (20 in.), ♀ 60 cm (24 in.).

Field Characters A round-winged 'true' hawk with a longish multi-banded tail—a magnified Sparrow-Hawk in effect. *Above*, greyish brown, darker on crown and with a white supercilium. *Below*, white, narrowly cross-barred with black. Tail greyish with 3 or 4 broad black bands. Sexes alike: female much larger. In overhead soaring flight close-barred body and underwing, broad blunt wings, and longish fanned tail with black bands (subterminal band broadest) are suggestive pointers.

Status, Habitat, etc. Uncommon winter visitor to the area; foothills and up to 3000 m: oak and conifer forest. Keeps singly. Hunts from ambush on a leafy branch or commanding rock, pouncing on quarry coming out to feed in open. Strikes it on the ground or pursues it with speed and tenacity flat along the surface. The larger, more powerful female highly prized in falconry. Has the characteristic habit of shooting steeply up into a branch for alighting. Food: large birds such as pigeons, partridges, pheasants, and small mammals—hares etc. Call: a mewing squeal and chattering *giak-giak-giak*.

10 CRESTED GOSHAWK *Accipiter trivirgatus* (Temminck)

Size House Crow + ; length 40–45 cm (16–18 in.).

Field Characters A short-winged or 'true' hawk. *Above*, dark brown; forehead, crown and short nuchal crest blackish grey. Tail with 4 visible black bars. *Below*, white, broadly streaked on breast and barred on rest with rich rufous-brown. A conspicuous blackish mesial streak down white throat to breast. Sexes alike: female larger. On a casual sighting confusable with Blyth's Baza (5), q.v.

Status, Habitat, etc. Resident: duars and foothills up to 2000 m—fairly open deciduous and semi-evergreen forest. Hunting tactics as common to all broad-winged hawks (genus *Accipiter*), namely pouncing from ambush in leafy tree overlooking a clearing and striking unsuspecting prey emerging to feed. If unsuccessful, chases the quarry with rapid wing beats close along the ground, weaving deftly in and out of tree-trunks and other obstacles in the pursuit. Food: small mammals, and birds as large as junglefowl. Call: a shrill prolonged yelp.

11 **SPARROW-HAWK** *Accipiter nisus* (Linnaeus)
Size House Crow − ; length 30–35 cm (12–14 in.).
Field Characters A medium-sized round-winged hawk. *Above*, slaty grey, darkest (blackish) on head, palest (greyer) on rump. A white supercilium. Tail tipped whitish, with 4 broad black bands. *Below*, white to pale rufous, finely striated with rufous on chin, throat and upper breast, narrowly barred with rufous-brown on rest of underparts. Sexes nearly alike: female larger and paler; barring on underparts less rufous more grey-brown.
Status, Habitat, etc. Resident; breeding between 1400 and 3500 m; observed as high as 4600 m in spring and summer (E. Nepal): forest and well-wooded country. Descends to the foothills and adjacent plains in winter. Typically of the genus, hunts from ambush up in leafy trees. Glides along at great speed close to the ground, accelerated by spurts of rapid wing beats in pursuit of quarry. The larger female (*basha*) a favourite with falconers. Food: chiefly small birds. Call: *tiu-tiu-tititi* quickly uttered.

12 **BESRA SPARROW-HAWK** *Accipiter virgatus* (Temminck)
Size House Crow − ; length 30–35 cm (12–14 in.).
Field Characters A medium-sized short-winged hawk. Very similar to Sparrow-Hawk, q.v., and easily confused with it. Distinguished by the broad black mesial stripe on throat (absent in Sparrow-Hawk, narrow and grey in Shikra) and two faint moustachial streaks on either side of it. Female larger than male.
Status, Habitat, etc. Resident. Breeding between 1000 and 2000 m, descending in winter to the foothills and duars: broken forested country. Food: mainly small birds. Call unrecorded.

13 **LONGLEGGED BUZZARD** *Buteo rufinus* (Cretzschmar)   p. xvi
Size Pariah Kite; length 60 cm (24 in.). .

Field Characters Like a robustly built kite but with rounded (not forked) tail. Plumage extremely variable, ranging from dark brown through reddish brown and fulvous to pale sandy. Tail distinctive: creamy rufous to cinnamon, unbarred. Sexes alike. In overhead flight large white patches on black-tipped primaries, and rufous unbarred tail, suggestive. Owing to great variability the field identification of buzzards with certainty is difficult.

Status, Habitat, etc. Winter visitor between 1500 and 3700 m: conifer forest and open hillsides with boulders and sparse scrub. Flight: rather sluggish flapping and much sailing, the broad wings held horizontal and tail partly expanded. Pounces on prey from a lookout perch or from air, occasionally hovering cumbrously to locate quarry on ground. Much given to soaring aloft and circling on motionless wings for long periods. Food: carrion, rodents, birds, reptiles and other small animals. Call: a loud wailing mew.

14 UPLAND BUZZARD *Buteo hemilasius* Temminck & Schlegel
Size Kite + ; length 70 cm (28 in.).

Field Characters Slightly larger than Longlegged. Also extremely variable in plumage, but normally more greyish brown above. Some with large irregular brown spots on underparts, some all brown below. Barred tail the only suggestive clue: brown (not rufous), with a broad blackish subterminal band and seven narrow dark bars.

Status Uncertain. Evidently rare visitor; recorded at high elevations in Sikkim in winter.

15 BUZZARD *Buteo buteo* (Linnaeus)
Size Kite − ; length 50–60 cm (20–24 in.).

Field Characters Like other buzzards very variable, with dark and light colour phases often making field identification uncertain. (One phase with creamy brown head and patchy underwing confusingly like immature Brahminy Kite.) Tail greyish brown, sometimes tinged with rufous, with one broad blackish subterminal band and numerous narrow dark and light bars (cf. Longlegged Buzzard). In overhead flight the broad black-tipped wings show large whitish patches.

Status, Habitat, etc. Poorly known due to dubious identifications. Recorded in winter from Nepal between 275 and 2240 m; may occur in the eastern Himalayas.

16 FEATHERTOED HAWK-EAGLE
    Spizaetus nipalensis (Hodgson)   p. xvi
Size Kite + ; length 70 cm (28 in.).

Field Characters A slender, crested, forest-haunting eagle. Identified in flight-silhouette by the short rounded wings (as in *Accipiter* hawks) upturned at ends, and longish unexpanded tail. *Above*, brown with a long black nuchal crest. Rump barred with white. Tail dark brown broadly barred with grey, and with a broad black subterminal band. *Below*, throat and breast fulvous white boldly streaked with black. A prominent black median gular stripe flanked by black moustachial streaks. Rest of underparts brown: flanks and under tail-coverts barred with white. Sexes alike: female larger. In very old birds black streaks on breast replaced by broken white bars. Young birds very variable.

Status, Habitat, etc. Resident, between 600 and 3300 m in the eastern Himalayas. Bold and powerful and very fast and adroit on the wing, though usually hunts ground game such as junglefowl, pheasant and hare by pouncing on them from a leafy ambush on the edge of a forest clearing—in the manner of *Accipiter*s. Call: a shrill metallic whistle, *kee-kikik*, reminiscent of Green Magpie (200), q.v.

17 **BONELLI'S HAWK-EAGLE** *Hieraaetus fasciatus* (Vieillot)   p. 16
Size Kite + ; length 70 cm (28 in.).

Field Characters A rather slender but powerful uncrested raptor with broad rounded wings and relatively long tail. *Above*, dark umber brown. Tail dark grey above, whitish underneath with a broad black terminal band and several others narrower and fainter. *Below*, white to buff, streaked with blackish. Sexes alike: female larger.

In overhead flight, white body, blackish under wing-coverts, finely grey-barred black-tipped flight quills and longish tail with broad black subterminal band are leading pointers.

Young birds confusing: uniformly brown or rufous-buff on underparts, streaked with blackish. Tail narrowly barred and mottled, and without the black subterminal band.

Status, Habitat, etc. Resident? Uncommon: well-wooded country—foothills and up to 2400 m. A bold and active hunter of small mammals and large birds such as junglefowl and pheasants. Either pounces on quarry from ambush in tree or chases it with great dash and agile twists and turns. Soars aloft like most raptors. Call: a shrill creaking chatter *kie-kie-kikiki* rather like Feathertoed.

18 **RUFOUSBELLIED HAWK-EAGLE**
   Lophotriorchis kienerii (E. Geoffroy)
Size Kite ± ; length 50–60 cm (20–24 in.).

Field Characters A handsome slender eagle with a prominent occipital crest. *Above*, glossy black, including crest. *Below*, throat and upper breast white with long black streaks; rest rufous-chestnut, streaked with black on belly and flanks. Sexes alike; female larger.

In flight, long wings and relatively short tail present falcon-like profile. A pale brown patch on upper side of wings and pale grey undersurface, white breast and rufous belly diagnostic.

Young. *Above*, brown with white line across forehead running back as supercilia. Tail blackish, broadly barred with grey. *Below*, white, sparsely black-streaked.

Status, Habitat, etc. Resident. Evergreen and moist-deciduous forested foothills up to 2400 m. Hunts large game birds and small mammals in the manner of hawk-eagles, qq.v. Call: a piercing plaintive scream.

19 BLACK EAGLE *Ictinaetus malayensis* (Temminck)   p. 16

Size Kite + ; length 70–80 cm (28–32 in.).

Field Characters A large black eagle with wings reaching to end of tail at rest. In sailing flight, exceptionally broad wings held in a flat V above line of back, splayed finger-like and upturned at ends. A pale patch on dark underside of wings, narrowly grey-barred tail, and bright yellow cere and feet conspicuous. Sexes alike: female larger.

Young. *Above*, pale brown stippled paler on head and hindneck. Upper tail-coverts fringed with white. *Below*, throat and breast with fulvous-brown oval drops; belly and flanks dark-streaked.

Status, Habitat, etc. Resident. Widely but patchily distributed in evergreen and moist-deciduous forested foothills, and up to 2700 m. Usually seen gliding lazily on broad outstretched wings close above the forest canopy, expertly weaving in and out among the upper branches and boles. Food: largely birds' eggs plundered from treetop nests; also large insects, frogs, lizards, small birds and rodents. Call: a subdued *kip*, *kip*, *kip* slowly uttered during a pair's aerial interplay. Normally silent.

20 RINGTAILED or PALLAS'S FISHING EAGLE

Haliaeetus leucoryphus (Pallas)

Size Kite + ; length 80 cm (32 in.).

Field Characters A heavy dark brown eagle with pale golden brown head and neck, and a broad white subterminal band in tail (conspicuous in flight). Sexes alike: female larger.

In sailing flight wings held in same plane as back, the tips slightly downcurved. Tail seldom spread out except in banking.

Young, dark brown; ear-coverts darker; wing-quills and tail (unbanded) blackish. Underparts paler and greyer.

Status, Habitat, etc. Resident: breeding up to 1800 m between October and February, part of the population migrating to Tibet etc. April to September. Affects rivers, lakes and upland bogs. Hunts fish from lookout mounds or trees by hurling itself on one near the surface, not plunging like Osprey. Also takes reptiles, frogs and waterfowl, but lives largely by piracy, bullying and robbing weaker raptors of their lawful prize. Call: loud creaking screams as of an unoiled wooden block tackle of a village well.

21 GREYHEADED FISHING EAGLE *Ichthyophaga nana* (Blyth) p. xvi
Size Kite ± ; length 60 cm (24 in.).
Field Characters *Above*, head and neck grey; rest dark brown. Tail: basal two-thirds mottled brown and white (looking greyish), terminal third blackish; central pair of rectrices all brown with a broad blackish subterminal band and pale tip. *Below*, breast pale brown; belly and flanks white. Sexes alike.

In flight, grey head, white belly and vent, and shortish grey-and-black tail diagnostic.

Young. *Above*, head and neck brown, streaked whitish; rest dark brown, the feathers edged paler. Wing-quills barred. *Below*, sides of head, chin, and throat greyish; breast and flanks pale brown broadly white-streaked. White of belly, flanks and tail heavily mottled with brown.

Status, Habitat, etc. Resident. The duars and foothills from under 1000 m up to at least 2000 m. Affects clear rapid forest streams. Predominantly fish-eating. Rather sluggish. Sits upright in branches overhanging water. Captures fish near surface by pouncing from the perch or swooping from the air and snatching it while quartering a likely section of the stream. Food: chiefly fish; occasionally birds and small mammals. Call: a loud clanging scream, uttered singly or in series, on the wing or while perched.

22 BLACK or KING VULTURE *Sarcogyps calvus* (Scopoli)
Size Peacock minus the train: length 80 cm (32 in.).
Field Characters A huge black turkey-like vulture, readily identified by naked scarlet head, neck, thighs and legs. White downy patches at base of neck and on upper thighs, additional pointers. Sexes alike.

In overhead aspect a whitish band on underside of rather pointed wings, and the white patches on thighs and upper breast diagnostic even from great height in the sky.

Young. *Above*, brown, the feathers edged paler. Crown covered with white down; neck partly feathered. *Below*, paler brown. Crop, lower belly and vent white.

Status, Habitat, etc. Resident but wide-ranging in search of food; locally up to 2000 m, in open country. Usually one or two birds among vulture gatherings at animal carcasses or soaring over valleys. Contrary to popular notion, rather timid and retiring. Food: carrion. Call: hoarse croaks when bickering at a carcass. Normally silent.

23 HIMALAYAN GRIFFON VULTURE *Gyps himalayensis* Hume

Size Enormous: length 125 cm (50 in.).

Field Characters The largest bird in the Himalayas: sandy white or pale khaki with whitish down-covered head and neck and pale brown ruff. Flight- and tail-quills black. *Below*, pale brownish buff broadly streaked with whitish. Sexes alike.

In overhead aspect large size, buff underside, black tail, and black trailing edges to the broad 'splayed-finger' wings diagnostic.

Young. Darker brown and chocolate-brown with broader whitish shaft-stripes on underparts.

Status, Habitat, etc. Purely a mountain bird. Resident, normally between 600 and 2500 m, wandering widely in search of food, up to 4500 m or higher. Affects open montane country. Sails majestically over mountaintop and valley scouting for animal carcasses. Gathers in numbers to feed at one with much squabbling and noise. Food: carrion. Call: hoarse croaks and screeches when scrimmaging at the feast. Normally silent.

24 BEARDED VULTURE or LÄMMERGEIER

Gypaetus barbatus (Linnaeus)   p. xvi

Size Equally large but less massive than Himalayan Griffon, and with longer tail. Length 125 cm (50 in.).

Field Characters Rather eagle-like, with cream-coloured fully feathered head and neck, and longish wedge-shaped tail. *Above*, silvery grey-and-black with bold white streaks. *Below*, rusty white. A beard of black bristle-like feathers at chin, conspicuous in profile. Sexes alike.

In overhead flight long, comparatively narrow wings and longish wedge-shaped tail readily identify it.

Young. Very dark brown with blackish head, and black goatee as in adult.

Status, Habitat, etc. Resident between 1200 and 4000 m, rarely (in winter) lower. Common but not abundant, ranging widely in search of food. Usually solitary. Sails majestically over mountain slopes and valleys, and round precipitous contours, often soaring to immense heights. Best known for its habit of carrying large bones aloft and dropping them on rocks below, then descending to feed on the splinters.

Has favourite bone-dropping places or ossuaries. Food: chiefly carrion.
Call: a sharp guttural *koolik, koolik* uttered in aerial courtship interplay.
Normally silent.

25 HEN HARRIER *Circus cyaneus* (Linnaeus)

Size Kite − ; slimmer. Length 45–50 cm (18–20 in.).

Field Characters A slender, graceful raptor normally seen quartering the
ground, gliding effortlessly on outstretched narrow wings, rising and
falling to the undulations. Male. *Above*, head, mantle, back and tail
bluish grey. Upper tail-coverts pure white. Primaries black. *Below*,
chin, throat and upper breast like back; rest white.

In flight black wing-tips contrasting with blue-grey plumage, and the
glistening white rump-patch, diagnostic pointers.

Female. Indistinguishable from female Montagu's Harrier
(*C. pygargus*). *Above*, dark brown with white rump-patch and dark-
barred tail. *Below*, pale rufous-brown with dark streaks. A well-
developed buff collar or ruff.

Status, Habitat, etc. Uncommon. Winter visitor and/or transient on
spring and autumn passage. Foothills and up to 2500 m: open hillsides
and grassy undulating valleys. Food: lizards, small birds, rodents, etc.
Call: Unrecorded. Very silent in winter.

26 PIED HARRIER *Circus melanoleucos* (Pennant)   p. xvi

Size Kite − ; slimmer. Length 45–50 cm (18–20 in.).

Field Characters Male. A slender black-and-white hawk. Head, mantle,
throat and breast black. Rest of underparts, and rump, white. Tail
grey. Wings grey, with a black band across upper surface (median
coverts) and broad black tip (primaries) conspicuous in flight. Female.
Dark brown above, with whitish rump and some white on nape; rufous
below. Indistinguishable from other female harriers without much
practice.

Young. Like adult female but with rufous edges to the head and
neck feathers; whitish nuchal patch prominent.

Status, Habitat, etc. Winter visitor to the duars and lower foothills.
Uncommon. Recorded (on passage?) at 1200 m; possibly occurs higher
during spring and autumn migration.

27 CRESTED SERPENT EAGLE *Spilornis cheela* (Latham)   p. xvi

Size Kite + ; length 70 cm (28 in.).

Field Characters A large dark brown eagle with full, rounded, black-and-
white nuchal crest, very prominent when erected. Underparts paler
brown, ocellated and finely barred with white and blackish. A bright

yellow patch at base of bill, and bright yellow unfeathered legs. Coloration variable, but sexes alike.

In sailing flight the broad rounded wings held in same plane as back. Underwing pattern striking: a narrow black border with a broad parallel white band across primaries along entire trailing edge, followed by narrower concentric black-and-white parallel bands. Tail black, with a broad whitish band posteriorly. Distinctive 3- or 4-noted shrill screams confirm its identity even at enormous heights in the sky.

Status, Habitat, etc.   Resident. Duars, foothills and up to 2300 m: jungle-clad ravines and forest edge; singly or in pairs. Sits bolt upright on a leafy bough commanding a clear view, and pounces on quarry on ground. Much given to soaring and circling aloft. Food: snakes, lizards, frogs, rodents, etc. Call: loud, high-pitched, prolonged whistling screams *kee-kee-keee* or *kek-kek-kek-keee* while soaring (sometimes at rest) usually prefaced by three short quick undertone notes like *pu-pu-pu*.

28 OSPREY *Pandion haliaetus* (Linnaeus)

Size Kite − ; length 60 cm (24 in.).

Field Characters A water-frequenting fish-eating raptor. *Above*, dark brown with white-streaked partially tufted head, and a prominent black band running from eye to nape. *Below*, pure white with a conspicuous brownish gorget across upper breast. Sexes alike: female larger.

In overhead aspect white underside with brown breast-band, closely barred pointed wings with black patches at the 'wrists' and barred squarish tail, and habit of hovering over water, diagnostic.

Status, Habitat, etc.   Mainly winter visitor, September to March. Breeds (rarely) between 2000 and 3300 m. Affects broad rivers, dammed reservoirs, lakes and tarns. Keeps singly perched on trees or stacks, or quartering the waterspread 20 or 30 m above with slow deliberate wing beats and gliding. Stops in mid air now and again and hovers cumbrously to investigate, legs dangling in readiness. Closes wings and hurls itself on quarry, sometimes submerging completely. Food: exclusively fish. Call: seldom heard in winter.

FALCONS: Falconidae

29 HIMALAYAN REDBREASTED FALCONET
Microhierax caerulescens (Linnaeus)   p. 16

Size Bulbul ± ; length 18 cm (7 in.).

Field Characters A diminutive falcon, very like a shrike when at rest. *Above*, glossy black. Face white with a prominent black band through eye. A broad white collar on hindneck. Tail square-ended. *Below*, chin, throat, thighs and under tail-coverts deep ferruginous; rest rusty white. Sexes alike: female larger.

In flight—rapid flaps of the pointed wings punctuated by graceful glides, tail partly spread—reminiscent of Ashy Swallow-Shrike (191).

Status, Habitat, etc. Resident: duars and foothills normally up to 650 m, rarely to 2000 m: outskirts of moist-deciduous and evergreen forest and abandoned cultivation with tall relict trees to serve as foraging bases. Pairs or family parties. Sits upright with shrike-like profile on bare branches, swinging tail slowly up and down like Blackwinged Kite. Swoops to the ground to whisk off prey in its talons or chases and captures it in mid air, circling back gracefully to the base with it. Actions on wing very like swallow-shrike, including the steep upward glide before alighting. Food: large insects: butterflies, dragonflies, grasshoppers, etc. Call: not recorded.

30 SHAHEEN FALCON *Falco peregrinus peregrinator* Sundevall
Size Jungle Crow ± ; length 40–45 cm (16–18 in.).

Field Characters A powerful, broad-shouldered, streamlined falcon with long pointed wings and swift direct flight. *Above*, slaty black with black head and prominent moustachial stripes on either side of throat. *Below*, pinkish white and rusty, boldly cross-barred with black from belly down. Sexes alike; female larger.

In overhead flight the bullet-shaped body, pinkish breast, narrowly barred belly and underside of the tapered wings are suggestive clues.

Status, Habitat, etc. Resident, possibly breeding on suitable crags up to 3000 m. Affects rather open, rugged montane country. Pairs occupy favourite crags all year and almost traditionally, as foraging bases over vast surrounding tracts. Flight very swift and direct—a few rapid wing beats followed by a long glide at great velocity. Hunts by lightning stoops from a 'pitch' aloft, binding the quarry in mid air. Pairs indulge in spectacular aerobatics during nuptial display, nose-diving through space and zooming up again with breath-taking abandon. Food: chiefly birds—partridges, pigeons, etc. Call: undescribed.

31 INDIAN HOBBY *Falco severus* Horsfield
p. 32

Size Pigeon ± ; length 30 cm (12 in.).

Field Characters A small, pointed-winged, streamlined falcon. *Above*, slaty grey with blackish head and moustachial streaks. *Below*, breast and underparts ferruginous. Sexes alike; female larger.

In overall effect a passable miniature of the Shaheen Falcon, q.v., but for absence of any blackish barring on underparts.

Status, Habitat, etc. Resident; breeding between 1800 and 2400 m, descending to the well-wooded foothills and duars in winter. Often seen in loose flocks of 10 or 12 birds hawking winged insects in forest clearings in the manner of swallows and swifts, dashing swiftly in pursuit, wheeling, circling, rising and falling in the air. Markedly crepuscular, hunting chiefly at dawn and dusk. Food: large insects—locusts, cicadas, dragonflies, etc.; occasionally lizards, small birds, small bats and mice. Call: a shrill rapid squeal *ki-ki-ki-ki*.

32 KESTREL *Falco tinnunculus* Linnaeus   p. 32
Size Pigeon ± ; length 35 cm (14 in.).
Field Characters A small, slender falcon with pointed wings and slightly
     graduated tail. Male. *Above*, brick-red with pear-shaped black spots,
     and ashy grey head. *Below*, buff, spotted with brown arrowhead marks.
     Female, rufous above (including head) cross-barred with blackish.
     Paler below, with the markings denser and more streak-like.
       In flight, pointed black wings, rounded grey tail with broad black
     terminal band, and habit of hovering—'hanging in space'—are leading
     clues.
Status, Habitat, etc.   Chiefly winter visitor; a small population resident
     and possibly breeding in upland valleys. In winter usually met with
     between 700 and 5500 m in open unforested country—cultivated
     uplands, sparsely scrubbed hillsides, and mountain passes above
     tree-line. Found singly, perched on bush or rock on the look-out for
     crawling prey, or quartering its feeding territory 30 metres or so up in
     the air. Stops dead in its flight every now and again and remains
     suspended in mid air with rapidly vibrating wing-tips to scan the ground
     below. On detection of movement drops a step lower, and so on, finally
     pouncing silently on the quarry and bearing it away in its claws. Food:
     large insects, lizards, small birds, mice, etc. Call: a shrill *kik-kik-kik*
     or *tit...wee* uttered during spectacular nuptial aerobatics.

PHEASANTS, JUNGLEFOWL, PARTRIDGES, QUAILS: Phasianidae

33 SNOW PARTRIDGE *Lerwa lerwa* (Hodgson)   p. 17
Size Partridge + ; length 40 cm (16 in.).
Field Characters A gregarious high-altitude partridge with prominent
     bright red bill and legs. *Above*, narrowly barred black and white,
     including head and neck. *Below*, from breast down, chestnut broadly
     streaked with white.
Status, Habitat, etc.   Resident between 3000 and 5000 m, descending only

a little lower in severe winters. Locally common. Affects alpine pastures and open hillsides above tree-line, with lichens, ferns and dwarf rhododendron and scattered snow patches. Family parties, or coveys of up to 20, scratch and feed on ground. When flushed, they rise with loud whirring and clapping of wings and scatter, flying off fast and strongly. Usually very tame, permitting close approach. Food: lichen, moss, seeds and shoots, with a quantity of gravel added. Call: in breeding season ringing musical chuckles similar to call of Grey Partridge of the plains. Also a shrill alarm whistle.

34 TIBETAN SNOWCOCK *Tetraogallus tibetanus* Gould   p. 33

Size Large domestic hen + ; length 70 cm (28 in.).

Field Characters A stout dumpy game bird reminiscent of a giant Grey Partridge. Male. *Above*, crown, nape and hindneck dark grey; a rufous band from gape down sides of neck. Upper back sandy grey; mantle blackish grey and rufous, finely streaked and vermiculated with buff. Rump rufous; tail rufous and blackish with rufous tip. A white wing-patch (secondaries) conspicuous in flight. *Below*, throat and upper breast white; a broad grey breast-band. Rest of underparts white broadly streaked with black. Female similar but breast-band vermiculated with fulvous.

The allied West Himalayan species, *T. himalayensis*, differs in having a chestnut 'necklace', and lower breast and belly dark grey.

Status, Habitat, etc. Resident in the high Tibetan plateau facies between 3500 and 5500 m—bare rocky, sparsely scrubbed hillsides and ridges, and alpine pastures above tree-line. Parties of 3 to 5 or larger coveys. When alarmed runs uphill with a waddling gait, jerking tail now and again to flash white under tail-coverts like moorhen. Mounts a rock before take-off and glides long distances effortlessly across valleys and from ridge to ridge. Food: bulbs, tubers, grass blades, etc. with much grit. Call: wild penetrating whistle of several notes (usually 5), frequently repeated, recalling both Green Pigeon and Curlew.

35 TIBETAN PARTRIDGE *Perdix hodgsoniae* (Hodgson)   p. 32

Size Grey Partridge; length 30 cm (12 in.).

Field Characters A high-altitude partridge reminiscent of the familiar Grey of the plains. *Above*, head distinctly patterned: white forehead running back as a prominent supercilium; ear-coverts chestnut with a black cheek-patch below. A dull chestnut collar at base of hindneck. Back buffy grey barred with blackish; tail largely chestnut. *Below*, white, barred with black on breast, more broadly with chestnut on flanks. Sexes alike.

Plate 2, artist Paul Barruel

OWLS, EAGLES, FALCONET

Inches
0   4   8

P.BARRUEL

P.BARRUEL

Plate 3, artist Paul Barruel

TRAGOPAN, PARTRIDGES, PHEASANT, GOOSANDER, IBISBILL

1 R SATYR TRAGOPAN, *Tragopan satyra*   page 20
Domestic fowl. ♀ Rufous-buff, wing-shoulder tinged
with crimson. Underparts with whitish spots. 2400–
4300 m.

2 R COMMON HILL PARTRIDGE, *Arborophila*
torqueola   18
Partridge − . ♀ Crown black-streaked brown. Throat
rufous, breast brownish, separated by chestnut band.
1500–4000 m.

3 R SNOW PARTRIDGE, *Lerwa lerwa*   14
Partridge + . Close-barred. Bright red bill and legs.
3000–5000 m.

4 R BLOOD PHEASANT, *Ithaginis cruentus*   19
Domestic fowl. ♀ Vermiculated rufous-brown. Face
and throat bright ochre; crest and nape grey. 2500–
4500 m.

5 R GOOSANDER or MERGANSER, *Mergus merganser*   1
Duck + . ♀ Head cinnamon-brown. Back grey-brown.
Throat white. Wings black-and-white. Thin bright
red bill, and legs as ♂. Up to 3000 m.

6 R IBISBILL, *Ibidorhyncha struthersii*   29
Grey Partridge + . Black mask; double breast-band.
Curved, curlew-like, red bill. Foothills to 4400 m.

Status, Habitat, etc. Resident, locally common, in the high Tibetan Plateau facies between 3500 and 5500 m, somewhat lower in winter: rocky hillsides with scattered furze bushes, and dwarf juniper and rhododendron scrub. In coveys of 10 to 15 which explode with whirr of wings on sudden alarm, the birds scattering and gliding downhill along the contours. Normally prefers running to flying. Food: seeds, shoots, roots, and presumably insects. Call: a rattling *sherrek-sherrek* etc. A shrill creaking while going off.

36 COMMON HILL PARTRIDGE

Arborophila torqueola (Valenciennes)   p. 17

Size Grey Partridge – ; length 30 cm (12 in.).

Field Characters A dumpy olive-brown short-tailed partridge with chestnut flanks. Male. *Above*, crown and nape bright chestnut with black spots on latter. *Below*, chin and throat black separated from grey breast by a white band. Rest of underparts grey and white with broad chestnut streaks on flanks. Female similar but crown brown, streaked with black. Breast brownish; throat-band chestnut instead of white.

Status, Habitat, etc. Resident between 1500 and 4000 m: ravines and hillsides clad in dense forest of oak etc. with thick evergreen undergrowth. Gregarious. Coveys of 5 to 10 scratch the mulch on the forest floor to feed. Normally trusts to legs for escape. When suddenly alarmed, flies strongly through the tree-trunks, often taking refuge up in leafy branches. Food: seeds, shoots, berries, insects, molluscs. Call: a single low mournful whistle repeated slowly two or three times followed by a series of 3 to 6 mellow double whistles ascending in scale, rendered as *bobwhite, bobwhite, bobwhite* . . .

37 RUFOUSTHROATED HILL PARTRIDGE

Arborophila rufogularis (Blyth)   p. 33

Size Grey Partridge – ; length 25 cm (10 in.).

Field Characters A dumpy short-tailed olive-brown partridge like 36. Male. *Above*, forehead grey; crown olive-brown stippled with black; long supercilia greyish white. Sides of face white stippled with black; white moustachial streaks; dark brown ear-coverts. *Below*, chin, throat and sides of neck rufous, spotted with black. Breast grey, separated from rufous throat by a ferruginous band and a black line below it. Flanks slaty grey and brown, broadly streaked with chestnut, with white drops and black crescentic spots. Female, more or less similar.

Status, Habitat, etc. Resident normally in a lower altitudinal zone than Common Hill Partridge (36)—between 1000 and 2400 m, descending slightly lower in winter: dense secondary scrub on abandoned cultivation clearings, and heavy undergrowth in evergreen forest. Habits similar to 36, q.v. Coveys of 6 to 12. An inveterate runner, difficult to flush without a dog. F o o d: weed seeds, fallen berries, shoots, insects, grubs, molluscs. C a l l: described as a clear loud double whistle *wheea-whu*, constantly repeated in ascending scale. But this needs verification.

38 WHITECHEEKED HILL PARTRIDGE
Arborophila atrogularis (Blyth)   p. 32
Size Grey Partridge − ; length 30 cm (12 in.).
Field Characters A typical dumpy olive-brown hill partridge like 36 and 37, easily distinguished from them by absence of any chestnut on flanks. Chin, throat and foreneck black, diffusing through black-and-white into grey breast and flanks; centre of belly whitish. Sexes alike.
Status, Habitat, etc. Resident, chiefly in Arunachal Pradesh (Dafla, Miri, Mishmi hills), from 1500 m down in the foothills almost to the duars: undergrowth in evergreen forest, as well as bamboo jungle. Gregarious, but feeds in widely scattered pairs or coveys of 5 to 8, the birds flushing one at a time when almost trampled on, giving the misleading impression of being solitary. F o o d: as in 36 and 37. C a l l: very soft mellow whistles to rally a dispersed covey. Other recorded descriptions need verification.

39 REDBREASTED HILL PARTRIDGE
Arborophila mandellii Hume   p. 33
Size Grey Partridge − ; length 30 cm (12 in.).
Field Characters A typical dumpy olive-brown hill partridge, like 36. Male. *Above*, crown and nape dull chestnut-brown; supercilia dark grey running back to meet on hindneck; lower neck and upper back chestnut with black spots. Rest of upper plumage olive, spotted and scalloped with black. *Below*, chin and throat olive-chestnut, separated from deep chestnut upper breast by a prominent double gorget of white and black. Rest of lower plumage grey, with chestnut and white markings on flanks. Sexes alike.
Status, Habitat, etc. Resident in an altitudinal zone between 350 and 2450 m, possibly higher: dense undergrowth in secondary moist forest. Habits and food as in other *Arborophila*s, qq.v. C a l l: a long-drawn *quoick* followed by a series of ascending double notes. But this needs verification.

40 BLOOD PHEASANT *Ithaginis cruentus* (Hardwicke)   p. 17
Size Domestic fowl; length 45 cm (18 in.).

Field Characters Large, stocky, partridge-like. Overall grey and apple-green streaked with yellow, with a bushy mop-like crest and bright red naked patch round eye bordered with black. Forehead black, chin and throat crimson. Upper breast splashed with crimson (like blood stains). Crimson patches on shoulders and under tail-coverts, and in tail. **Female.** Strikingly different: overall rufous brown, finely vermiculated. Forehead, face and throat bright ochre contrasting strongly with grey crest and nape.

Status, Habitat, etc. Resident, in three slightly differing subspecies, between 2500 and 4500 m, moving up and down with the snow-line, and somewhat lower in severe winters. Affects conifer forest, dwarf rhododendron, dense ringal bamboo and juniper scrub interspersed with snow patches. Coveys of 5 to 10, sometimes more. Usually tame and confiding. Feeds in open patches, often digging with the stout bill in deep snow. Food: mosses, ferns, pine shoots, lichens, etc. Call: a prolonged high-pitched squeal to rally a scattered covey.

41 SATYR TRAGOPAN *Tragopan satyra* (Linnaeus)   p. 17

Size Domestic fowl; length ♂ 70 cm (28 in.), ♀ 60 cm (24 in.).

Field Characters. Male. *Above*, head black with a crimson streak or 'horn' on either side of recumbent crest. Rest of upperparts crimson and brown, sprinkled with black-bordered white spots or ocelli. *Below*, a naked deep blue patch or lappet on throat, bordered with black. Rest of underparts crimson, sprinkled with white ocelli as above. **Female.** *Above*, rufous-buff, vermiculated, barred and blotched with black, narrowly streaked with pale ochre on crown. Wing-shoulder tinged with crimson. Tail rufous-brown barred with black and buff. *Below*, chin and throat whitish, breast pale rufous-buff; belly still paler and with large whitish spots.

Status, Habitat, etc. Resident between 2400 and 4300 m, descending to lower levels in severe winters. Keeps to khuds and steep hillsides with scrubby undergrowth and ringal bamboo, in oak, deodar and rhododendron forest. Pairs or small parties, scratching and feeding like junglefowl. Food: mostly ferns and other leafy vegetable matter. Call: a loud *wak* repeated several times. An arresting *kya ... kya ... kya ...* like the bleating of a goat kid.

42 GREYBELLIED TRAGOPAN *Tragopan blythii molesworthi* Baker

Size Domestic fowl; ♂ 70 cm (28 in.), ♀ smaller.

Field Characters Male. *Above*, forehead, crown, a patch on either side of neck, and border to the orange-yellow face and gular patch, black. Broad supercilia, hindneck and upper back bright crimson. Rest of upperparts brown tinged with crimson and sprinkled with white and maroon ocelli. *Below*, upper breast bright crimson forming a narrow gorget; lower breast and belly pale smoky grey, the feathers slightly paler-centred. Flanks and thighs mottled black and buff, splashed with crimson as on vent. Female. Rather like ♀ Satyr, q.v.; tail more rufous. *Below*, mottled, stippled and spotted with dark brown, rufous and greyish white; centre of belly and vent more uniform grey.

Status, Habitat, etc.   Little known. Resident, E. Bhutan and Arunachal east to the Mishmi Hills, between 1800 and 3300 m: thick undergrowth of shrubs and ringal bamboo on hillsides, in rhododendron and other high forest. Singly or in pairs; sometimes small parties. Scratches and feeds on the ground like junglefowl, and like them also roosts up in trees. Exceedingly shy and wild, and a great skulker, seldom giving a clear glimpse of itself. Food: as in the Satyr. Call: unrecorded.

43 TEMMINCK'S TRAGOPAN

Tragopan temminckii (J. E. Gray)   p. 32

Size Domestic fowl; length 60 cm (24 in.).

Field Characters Male. Overall bright crimson, with round black-bordered pearly grey spots or ocelli on upperparts. Head, borders of naked cobalt-blue face, and loop round similar gular-patch (lappet) black. Large triangular (or diamond-shaped) pearly grey spots on underparts. Two blue horns erected and brilliant blue-and-red throat lappet expanded during nuptial display. Female. *Above*, rufous to greyish brown, mottled with black and with whitish arrow-shaped marks. Tail like back with the markings forming dark and light bands. *Below*, chin and throat whitish with black lines. Rest of underparts pale brown with broad whitish shaft-spots and black blotches.

Status, Habitat, etc.   Resident. Luhit Frontier Division of Arunachal Pradesh (Mishmi Hills, Dibang and Tsangpo valleys, etc.) between 2100 and 3500 m: dense undergrowth on steep hillsides in moist ever-green forest. Usually solitary, never more than two or three together. General habits and food as of other tragopans, but is apparently more arboreal. Call: unrecorded.

44 MONAL PHEASANT *Lophophorus impejanus* (Latham)   p. 33

Size Large domestic hen + ; length 70 cm (28 in.).

Field Characters A dumpy brilliantly coloured pheasant. Male. *Above*, lustrous metallic bronze-green, purple and blue, with a large white rump-patch and upstanding crest of wire-like spatula-tipped feathers. Tail cinnamon-rufous, short, broad, square-cut. *Below*, velvety jet black. When flushed and flying off, the wild ringing squeals, chestnut wings and tail, and broad white rump-patch diagnostic. Female, brown, mottled and streaked with paler and darker brown, with a short tuft on crown. A white throat-patch and naked blue skin round eye conspicuous.

Status, Habitat, etc. Resident between 2500 and 5000 m: high forest of oak, rhododendron, deodar, etc. interspersed with glades and pastures and steep scrubby hillsides. Singly, pairs or parties of 3 or 4, digging vigorously with the powerful bill into the soil or snow for food. If surprised in the open invariably flies downhill; if in forest takes refuge in trees, sitting motionless and becoming remarkably invisible. Food: seeds, roots, tubers, berries, insects. Call: a wild, ringing whistle reminiscent of the Curlew's.

45 SCLATER'S MONAL PHEASANT *Lophophorus sclateri* Jerdon

p. 33

Size Same as the Monal (44).

Field Characters Very like 44, q.v.; somewhat duller and less refulgent, and with tuft of short curly feathers instead of upstanding crest. Tail cinnamon with a broad white terminal band. Crestless head and white-ended tail diagnostic. Female. Like 44 but with conspicuous greyish white rump. Tail blackish, tipped with white and with some narrow whitish bars.

Status, Habitat, etc. Resident. Extreme eastern Bhutan (?), and Arunachal Pradesh in the Kameng, Subansiri, Siang and Luhit frontier divisions, between 3000 and 4000 m: silver fir forest with dense rhododendron undergrowth. Food: as in 44. Call: a ringing whistle like 44 but distinct in tone.

46 ELWES'S EARED PHEASANT

Crossoptilon crossoptilon harmani Elwes

Size Large domestic fowl + ; about that of Monal. Length 70 cm (28 in.).

Field Characters A stout ashy grey pheasant with laterally compressed metallic blue-black tail with long, arching central feathers. Head velvety black with two white ear-tufts projecting like horns behind nape. Sides of head naked deep scarlet. *Below*, chin and throat white, running up into the white nape-band. Along middle of foreneck and belly white; rest of underparts deep ashy grey. Female similar but smaller.

Status, Habitat, etc. Resident; plentiful wherever occurring. Extreme northern fringe of Arunachal Pradesh bordering Tibet, normally between 3000 and 5000 m, rarely lower: grassy clearings on hillslopes in dwarf rhododendron and juniper scrub. Flocks of 5 to 10. Takes refuge in bushes when disturbed and is loath to fly. If flushed with a dog, flies up into a tree if available, otherwise coasts downhill for long distances. Food: shoots, tubers, berries, insects. Call: loud, harsh, distinctive, reminiscent of guineafowl; also a single squawk like heron's.

47 **KALEEJ PHEASANT** *Lophura leucomelana* (Latham)   p. 33
Size Domestic fowl ± ; length 60–70 cm (24–28 in.).
Field Characters Male. *Above*, glossy blue-black; a long, pointed, recumbent crest and naked bright scarlet patches round eyes. Feathers of lower back and rump with narrow white edges producing a scalloped or scaly pattern. Long, black, arching sickle-shaped tail. *Below*, chin, throat, and foreneck glossy blackish with bold white streaks. Feathers of breast and flanks lanceolate, dull brown with whitish scalloping. Belly and vent brown. Female dark rich brown with narrow whitish scalloping; crest like male but brown and shorter; naked scarlet patches round eyes.
Status, Habitat, etc. Resident at fairly low elevations, between 100 m in the foothills and up to 2500 m: thickly overgrown steep gullies on hillslopes, especially in the proximity of running water and terraced cultivation. Three subspecies with slight plumage differences: black-backed *melanota* (Sikkim, W. Bhutan), black-breasted *lathami* (E. Bhutan, Arunachal Pradesh), and all-black *moffitti* (probably C. Bhutan, but provenance enigmatical). Pairs or small parties emerge to scratch and feed in forest clearings and jungle tracks, chiefly mornings and evenings. Families roost communally up in trees. Food: seeds, tubers, berries, green vegetable matter, insects. Call: when flushed a sharply repeated *kurchi, kurchi, kurchi* ... Cock gives a challenging crow in the breeding season.

48 **RED JUNGLEFOWL** *Gallus gallus* (Linnaeus)   p. 33
Size Domestic fowl; length ♂ 70 cm (28 in.), ♀ 45 cm (18 in.).
Field Characters Both sexes very similar to the domestic 'Game Bantam' breed. Male. *Above*, chiefly glossy deep orange-red with long lanceolate yellowish neck-hackles. Long, metallic black sickle-shaped tail. *Below*, dull blackish brown. Female. *Above*, crown dull rufous; forehead and supercilia bright chestnut, continuing downward to meet in a loop on foreneck. Rest of upperparts reddish brown, finely vermiculated with buff. *Below*, pale-shafted light rufous-brown.
Status, Habitat, etc. Resident; chiefly duars and foothills, but locally up to 2000 m: moist-deciduous and evergreen forest, and bamboo and scrub jungle in broken hummocky country, especially where interspersed with clearings and cultivation. In eastern Arunachal Pradesh (Mishmi

Hills) intrudes the Burmese subspecies *spadiceus* with blunter neck-hackles, deeper golden red near tip. Keeps in small parties, usually a cock and 3 or 4 hens. Habits very similar to Kaleej Pheasant (47) q.v., the two species often associating. Food: seeds, shoots and other vegetable matter, insects, grubs, etc. Call: crow of cock very similar to that of Game Bantam—somewhat shriller and with a more abrupt ending.

49 PEACOCK PHEASANT *Polyplectron bicalcaratum* (Linnaeus) p. 32
Size Domestic hen; length ♂ 60 cm (24 in.), ♀ 50 cm (20 in.).
Field Characters Male. *Above*, head and neck brownish buff with a short frowzled mop-like crest; naked facial skin yellowish pink. Rest of upper plumage greyish brown, spotted and barred with whitish on back, rump and upper tail-coverts. Mantle, wing-coverts, and feathers of broad fan-shaped tail studded with brilliant violet green-blue, white-rimmed, eye-spots or ocelli. *Below*, chin and throat whitish; rest greyish brown, narrowly barred with whitish. Female, similar to male but smaller and duller, with shorter tail. Whitish on throat more extensive; ocelli on mantle less brilliant, those on the shorter rectrices obsolete.
Status, Habitat, etc. Resident. Duars and foothills up to 1200 m, in dense evergreen forest. Skulks in thick undergrowth; extremely swift on its legs and loath to fly. Difficult to flush even with a dog, thus seldom seen in the wild. Food: grain, seeds, berries, grubs, insects—especially white ants. Call: a deep guttural *ok-kok-kok-kok* not unlike a cock English pheasant's.

BUSTARD-QUAILS: Turnicidae

50 YELLOWLEGGED BUTTON QUAIL *Turnix tanki* Blyth
Size Grey Quail ± ; length ♂ 15 cm (6 in.), ♀ slightly larger.
Field Characters Male. *Above*, crown blackish, rufous and buff with a pale median line or 'centre parting'; supercilia and sides of head buff. Back greyish brown cross-barred and vermiculated with black. *Below*, pale buff; whitish on throat, darker rusty brownish on middle of breast, black-spotted on its sides. Female similar but slightly larger and with a broad orange-rufous collar on upper back and sides of neck, conspicuous when bird flying away from observer. Both sexes readily distinguished from co-existing Bustard-Quail (51) by bright yellow bill and legs *v.* bluish slaty.
Status, Habitat, etc. Resident (or nomadic ?). Duars and foothills up to 1200 m; rarely to 2000 m: damp grassland intermixed with low

shrubbery. Pairs; sometimes small parties. A great skulker, walking away quietly through undergrowth on disturbance and difficult to flush. Flies low and silently and tumbles into undergrowth after a short flight. Female polyandrous. Food and Call: as in Bustard-Quail (51), q.v.

51 BUSTARD-QUAIL *Turnix suscitator* (Gmelin)
Size Grey Quail ± ; length ♂ 15 cm (6 in.), ♀ slightly larger.
Field Characters Male. *Above*, crown blackish with a faint median stripe ('centre parting'); sides of head white stippled with black. Rest of upper plumage variegated and spotted with rufous-brown, black, white and buff. *Below*, chin and throat whitish. Rest of lower parts rusty buff banded with black on foreneck and breast. Female slightly larger, and more richly coloured. *Above*, like male. *Below*, rusty buff: chin, throat, foreneck and centre of breast black; sides of breast and anterior flanks barred with black. Both sexes distinguished from Button Quail by barred underparts and bluish slaty (*v.* yellow) bill and legs. In flight prominent buff patches on wing-shoulders are good pointers to its identity.
Status, Habitat, etc. Resident (and seasonally nomadic ?). Duars and foothills, occasionally up to 2000 m: grassland and scrub jungle. Habits very similar to Button Quail, q.v. Food: grass- and weed-seeds, green shoots, ants, termites and other insects. Call: a loud drumming as of a 2-stroke motorcycle engine in the distance, uttered by female. A subdued, but far-carrying booming *hoon ... hoon ...* (sex ?) in runs of a few seconds' duration.

CRANES: Gruidae

52 BLACKNECKED CRANE *Grus nigricollis* Przevalski   p. 32
Size Vulture; long-legged, standing 150 cm (5 ft) to top of head.
Field Characters A typical ashy grey crane with black head and neck. Lores and entire crown naked and dull red (*v.* merely a nape patch in Common Crane). A small patch of white feathers below and behind eye. Wing-quills black, the inner secondaries elongated into arching plumes falling over and concealing the black tail. Sexes alike; female slightly smaller.
Status, Habitat, etc. Regular winter visitor in small numbers to Bhutan and Arunachal Pradesh (Subansiri frontier division)—recorded in the Bumthang and Apa Tani valleys, 1500–3000 m. Breeds in Tibetan Plateau country between 4300 and 4600 m. In the above wintering areas

affects fallow paddy fields and swampy land—sedentary flocks of 20 to 40. Food: grain, seeds, shoots, tubers, insects. Call: loud, high-pitched trumpeting of Sarus pattern in uneven chorus, accompanied by the typical prancing and capering.

RAILS, COOTS: Rallidae

53 ELWES'S CRAKE *Amaurornis bicolor* (Walden)
Size Grey Quail ± ; length 22 cm (8½ in.).
Field Characters A quail-like chestnut and grey marsh bird. *Above*, head and neck dark ashy grey; rest of upper plumage rufous-brown. *Below*, chin whitish; lower plumage dark ashy grey. Sexes alike.
Status, Habitat, etc. Resident. Duars and up to at least 2800 m (Bhutan); reported at 3600 m (Sikkim). Affects dense swampy jungle, often bordering paddy fields. A great skulker, emerging to feed at the edge of cover early mornings and evenings, scuttling in on the least alarm. Food: insects, molluscs, worms, seeds of marsh plants. Call: unrecorded.

54 WHITEBREASTED WATERHEN
   Amaurornis phoenicurus (Pennant)
Size Grey Partridge ± ; length 30 cm (12 in.).
Field Characters A slate-coloured stub-tailed swamp bird with prominent white facial mask and underparts. *Above*, forehead and face pure white; crown, hindneck, and rest of upper plumage dark slaty grey. *Below*, chin to vent pure white; sides of breast, and flanks grey; posterior flanks and under tail-coverts rufous. Sexes alike.
Status, Habitat, etc. Resident. Duars and foothills up to 1500 m: reedy and scrub-covered wetland, edges of flooded rice fields, and ponds and ditches. Constantly jerks up the stumpy tail while sauntering about in search of food, flashing the chestnut under tail-coverts. Clambers freely up into bamboo clumps and shrubs. Flight feeble and typically rail-like, with legs dangling. Swims buoyantly. Food: insects, molluscs, worms, shoots and seeds of marsh plants. Call: silent except in breeding season when very noisy. Loud, raucous croaks and chuckles followed by long runs (maybe 15 minutes or more) of a monotonous metallic *krr kwaak-kwaak, krr kwaak-kwaak*, etc., or just *kook . . kook . . kook . .* like Coppersmith barbet's, chiefly during night and on cloudy overcast days.

PLOVERS, SANDPIPERS, SNIPE: Charadriidae

55 PEEWIT or LAPWING *Vanellus vanellus* (Linnaeus)

Size Grey Partridge; length 30 cm (12 in.).

Field Characters A typical plover. Overall effect black and white with a long slender upstanding crest. *Above*, forehead, crown and crest glossy black; sides of head and neck white with black patches; back and mantle metallic bronze-green. Upper tail-coverts cinnamon; tail white with broad black terminal band. *Below*, throat, foreneck and breast black; under tail-coverts cinnamon. Rest of lower parts white. Sexes alike.

In overhead flight plumage looks pied. Black breast and white underparts, including underside of wings, suggestive diagnostic pointers.

Status, Habitat, etc. Winter visitor (October to March): erratic and sparse. Duars and adjoining low foothills—fallow land, stubbles, irrigated stubble fields, etc. Pairs or small flocks. Runs about with short mincing steps, stopping abruptly time and again to pick up titbits with steep downward tilt of body, as characteristic of plover family. Normal flight rather slow, with sluggish wing beats. Food: insects, molluscs, worms. Call: silent in winter; only occasional plaintive mewing *pee-wit* uttered singly.

56 REDWATTLED LAPWING *Vanellus indicus* (Boddaert)

Size Grey Partridge + ; length 35 cm (14 in.).

Field Characters The commonest plover. *Above*, head and neck black; a broad white band from behind eye running down sides of neck to meet the white underparts; a crimson fleshy wattle in front of each eye. *Below*, throat and breast black; rest white. Sexes alike. Identity self-proclaimed by the well known *Did-ye-do-it?* calls.

Status, Habitat, etc. Resident. Duars and foothills to at least 1000 m— wet fallows and stubbles, usually in the neighbourhood of water: ditches, rain puddles, etc. Pairs or small parties. Runs about and feeds like 55. Always exceedingly vigilant and first to give the alarm on approach of an intruder. Food: insects, molluscs, worms and some vegetable matter. Call: loud, high-pitched, penetrating *Did-ye-do-it?* or *Pity-to-do-it* reiterated with vehemence varying with the occasion.

57 SPURWINGED LAPWING *Vanellus spinosus* (Linnaeus)

Size Grey Partridge; length 30 cm (12 in.).

Field Characters *Above*, forehead, crown, and occipital crest black. Upper plumage chiefly pinkish grey and sandy brown; tail-coverts white; tail white with its terminal half black. Wings black with a broad white patch. *Below*, cheeks, chin and throat black, bordered with white. Rest of lower plumage brownish grey and white; centre of belly black. In flight, black wings with broad white band combined with the black-and-white head pattern and black belly-patch diagnostic. At rest, the hunchbacked posture with rigid horizontal body and furtively drawn-in neck is characteristic.

Status, Habitat, etc.   Resident. Chiefly duars—on sandbanks and shingle beds in the Teesta, Luhit and other rivers of the region. Coloration remarkably obliterative in its shingly environment. Singly or pairs. Habits similar to 55 and 56. F o o d : insects, worms, molluscs, crustaceans. C a l l : a sharp, insistent, high-pitched *did, did, did* several times repeated, ending with 2 or 3 staccato notes sounding like *did-did-do-weet*, *did-did-do-weet* uttered on the wing. Reminiscent of Redwattled Lapwing's, but quite distinct.

58 SOLITARY SNIPE *Capella solitaria* (Hodgson)

Size Grey Quail + ; length (including bill) 30 cm (12 in.).

Field Characters Typical snipe with long slender straight bill and concealing plumage pattern of variegated brown, black, rufous, fulvous and white. Large size, pale coloration, solitary habit and mountain habitat suggestive clues, but field identification usually not definitive. Sexes alike.

Status, Habitat, etc.   Resident, uncommon. Breeds locally between 2800 and 4600 m in summer; descends to the foothills and duars in winter. Affects sprawling boggy mountain streams, often partly frozen, and grassy bogs amid rhododendron and suchlike scrub. Usually solitary; sometimes 2 or 3 widely separated individuals on same bog. Flight comparatively slow and heavy, but with the characteristic rapid twists and turns. Seldom flies far before pitching into cover. F o o d : worms, grubs, tiny molluscs. C a l l : the characteristic *scape* or *pench* on flushing, but deeper and harsher than Fantail Snipe's.

59 WOOD SNIPE *Capella nemoricola* (Hodgson)

Size Grey Quail; length (including bill) 30 cm (12 in.).

Field Characters *Above*, dark brown, concealingly patterned with black, rufous, and buff streaks. *Below*, breast fulvous barred with brown; rest white, barred closer. Sexes alike. Doubtfully distinguishable from Fantail Snipe, but larger size, darker coloration, comparatively slow bat-like flight with downward pointing bill and more wooded, hilly habitat suggestive.

Status, Habitat, etc.   Resident between 1300 and 3700 m, lower in winter— swampy patches among tall grass and bushes in hilly country. Solitary

or scattered twos and threes. Flight slow and wavering, recalling Painted
Snipe and Woodcock. Food: worms, grubs, aquatic insects, molluscs.
Call: usually flushes silently; occasionally with a low croaking *tok-tok*.

60 **WOODCOCK** *Scolopax rusticola* Linnaeus
Size Partridge − ; length (including bill) 35 cm (14 in.).
Field Characters An outsized snipe of highly concealing plumage. *Above*,
brownish grey, blotched and barred with black, rufous, and buff.
Hindcrown, nape and rump cross-barred black and rufous. *Below*,
chin white; rest brownish white cross-barred with blackish. Sexes
alike. Large tubby shape, sluggish flapping flight with bill pointing
groundward, short scimitar-shaped wings, and short fanned tail
suggestive clues.
Status, Habitat, etc.   Resident, between 2000 and 3800 m, descending
lower in winter, then migrating (in part) to the SW. peninsular hills.
Affects swampy glades with shrubby undergrowth in rhododendron,
fir, and mixed forest. Solitary or separated pairs; largely crepuscular
and nocturnal. Feeds like snipe by probing into wet earth with the
sensitive-tipped bill. Wavering flight accompanied by swift turns and
twists to avoid tree-trunks etc., and fluttering, almost perpendicular
drop into cover, characteristic. Food: worms, grubs, molluscs; some
seeds. Call: usually silent. Sometimes a deep croak and shrill bat-like
squeaks when flying around at dusk.

STILTS, AVOCETS,
IBISBILL: Recurvirostridae

61 **IBISBILL** *Ibidorhyncha struthersii* Vigors   p. 17
Size Grey Partridge + ; length 40 cm (16 in.).
Field Characters An ashy grey-brown sandpiper with black facial mask
and breast-band, and long downcurved red bill. Sexes alike. In flight,
resembling Stone Curlew's, outstretched neck and long curved red bill
give truncated look to short-tailed body. White wing-patches (concealed
at rest), and black breast-band are additional pointers.
Status, Habitat, etc.   Resident, between 1700 and 4400 m, descending
partly to the foothills in winter. Affects shingle banks and islets in placid
reaches of clear fast-flowing streams. Pairs or small parties, 6 to 8.
Feeds like typical sandpiper, often wading in breast deep and ducking
head completely to thrust specialized bill under and around stones.
On alarm, bobs head and wags posterior very like Greenshank. Food:

insects, worms, molluscs, crustaceans. Call: loud ringing sandpiper-like single whistle quickly repeated in flight.

PIGEONS, DOVES: Columbidae

62 PINTAILED GREEN PIGEON *Treron apicauda* Blyth   p. 36
Size Pigeon ± ; length (including long pointed tail) 40 cm (16 in.).
Field Characters Overall yellowish green with long pin-pointed tail and two conspicuous yellow bars in the black wings. Male with broad olive-green collar on hindneck and orange-pink wash on breast. Female without either, and with shorter tail.
Status, Habitat, etc.   Resident, subject to local nomadism governed by fruiting seasons. Duars and foothills up to at least 2500 m. Exclusively frugivorous and arboreal, only rarely descending to ground to drink. Flocks of 10 to 30 clamber about and feed among fruit-laden twigs, their plumage perfectly camouflaged in the foliage. Food: drupes and berries; largely wild figs. Call: a range of mellow musical whistles similar to those of the Wedgetailed species (63), q.v.

63 WEDGETAILED GREEN PIGEON *Treron sphenura* (Vigors)
Size Pigeon; length 35 cm (14 in.).
Field Characters Yellowish green, similar to 62 but with tapering wedge-shaped tail instead of pin-pointed and parakeet-like. Male. *Above*, crown tinged with orange-rufous; mantle largely maroon. *Below*, chin and throat yellow, breast pale orange-pink. Female like male but without rufous crown or maroon mantle.
Status, Habitat, etc. Resident, subject to spatial and altitudinal nomadism depending on fruiting seasons: duars and foothills, up to 2500 m. Affects broad-leaved forest. Gregarious, frugivorous, arboreal. Flocks of 6 to 15 or so, sometimes in feeding association with Pintailed. Habits and food of the two very similar. Call: also similar—a series of mellow, musical, wandering whistles roughly rendered as *ko-kla-oi-oi-oilli-illio-kla*.

64 THICKBILLED GREEN PIGEON *Treron curvirostra* (Gmelin)
Size Pigeon – ; length 25 cm (10 in.).
Field Characters A small yellowish green pigeon with chestnut-maroon mantle, a prominent yellow wing-bar, and grey and olive-green tail. The thick bright red and greenish bill, and vivid green naked skin round eye are leading diagnostic pointers. Female lacks the chestnut on

mantle and has whitish under tail-coverts barred with dark green (*v.* uniform pale cinnamon in male).

Status, Habitat, etc. Resident, subject to local movements governed by fruit ripening. Duars and foothills up to 1500 m. Affects well-wooded country and broad-leaved forest. Gregarious and exclusively frugivorous but reported to be less entirely arboreal. Frequently descends to feed on wild strawberries and berries of other ground plants. F o o d : berries, drupes and (predominantly) wild figs. C a l l : mellow musical whistles similar to other green pigeons'. Not specifically described.

65 ASHYHEADED GREEN PIGEON *Treron pompadora* (Gmelin)

Size Pigeon − ; length 30 cm (12 in.).

Field Characters A small yellowish green pigeon with dark ashy grey crown and nape sharply demarcated from adjacent greenish yellow parts. Mantle chestnut-maroon. A broad yellow band in black wings. Tail black-and-green, with a broad ashy terminal band on all but central pair of feathers which wholly green. *Below*, chin and throat greenish yellow; breast conspicuously orange; under tail-coverts cinnamon. F e m a l e lacks chestnut mantle and orange breast; under tail-coverts buff, mottled with dark green. Distinguished from very similar female of Orangebreasted *T. bicincta* (66), by green (*v.* grey) central tail-feathers.

Status, Habitat, etc. Resident, subject to local nomadism dependent on supply and ripening of fruit. Duars and up to 1500 m: forest and well-wooded country; sometimes large flocks of over 200. Frugivorous, and arboreal, not differing markedly in habits or calls from other green pigeons.

66 ORANGEBREASTED GREEN PIGEON *Treron bicincta* (Jerdon)

Size Pigeon ± ; length 30 cm (12 in.).

Field Characters Similar to Ashyheaded (65), but lacks the grey crown and chestnut-maroon mantle. *Below*, yellowish green with a lilac band across upper breast followed by orange band across lower. Under tail-coverts cinnamon, the longest edged pale yellow. All tail-feathers (except central pair) slaty grey above with broad blackish subterminal band; black below with grey tip. Central pair wholly grey. F e m a l e lacks lilac and orange breast-bands; under tail-coverts pale crimson with greenish mottling. Differentiated from very similar female Ashyheaded (65) by middle tail-feathers being slaty grey instead of green.

Status, Habitat, etc. Resident, with the usual seasonal nomadism. Duars and foothills up to 1500 m: forest and well-wooded country.

32

0   6   12   18   24   30 cm

Plate 5, artist J. P. Irani

DOVES, PARTRIDGES, SNOWCOCK, JUNGLEFOWL, PHEASANTS

1 R RUFOUS TURTLE DOVE, *Streptopelia orientalis*   page 38
Pigeon. Black-and-grey 'chessboards' on neck.
Broadly grey-fringed tail. Up to 4000 m.

2 R MAROONBACKED IMPERIAL PIGEON, *Ducula badia*   34
Jungle Crow + . Mantle maroon-brown. Broad black
subterminal band in tail. Up to 2300 m.

3 R EMERALD DOVE, *Chalcophaps indica*   39
Myna + . Mantle brilliant bronze-green. White
forehead and eyebrows. White band across rump.
Up to 1800 m.

4 R REDBREASTED HILL PARTRIDGE, *Arborophila mandellii*   19
Grey Partridge − . Breast chestnut. Double black and
white gorget on throat. 350–2450 m.

5 R RUFOUSTHROATED HILL PARTRIDGE, *Arborophila rufogularis*   18
Grey Partridge − . Long whitish supercilia. Rufous
throat separated from grey breast by ferruginous
band and black line. 1000–2400 m.

6 R TIBETAN SNOWCOCK, *Tetraogallus tibetanus*   15
Domestic fowl + . White throat. Grey breast-band.
3500–5500 m.

7 R RED JUNGLEFOWL, *Gallus gallus*   23
Large domestic hen. ♂ Breast black. ♀ Breast rufous-
brown, pale streaked. Up to 2000 m.

8 R KALEEJ PHEASANT, *Lophura leucomelana*   23
Domestic fowl ± . ♂ Pointed crest. Bare red face.
Whitish band across rump. Up to 2500 m.

9 R MONAL PHEASANT, *Lophophorus impejanus*   21
Large domestic fowl + . ♂ Crest of spatula-tipped
wire-like feathers. Rufous tail. ♀ White throat-
patch. Bare blue skin round eye. 2500–5000 m.

10 R SCLATER'S MONAL PHEASANT, *Lophophorus sclateri*   22
Domestic fowl + . ♂ Crest of short curly feathers.
Tail rufous with broad white terminal band. ♀ Lower
back whitish with brown bars. 3000–4000 m.

Gregarious, frugivorous and arboreal. General habits, food and calls as in other green pigeons, qq.v.

67 MAROONBACKED IMPERIAL PIGEON
Ducula badia (Raffles)   p. 33

Size Jungle Crow + ; length 50 cm (20 in.).

Field Characters A large greyish brown forest pigeon with whitish under tail-coverts. *Above*, crown and hindneck lilac; back and mantle maroon-brown; rump ashy grey. Tail black and brownish grey. *Below*, sides of head grey; throat white; rest vinous grey, under tail-coverts buff. Under aspect of tail grey with a broad black subterminal band. Sexes alike.

Status, Habitat, etc. Resident, with seasonal nomadism as in all fruit pigeons. Duars, foothills and up to 2300 m: tall evergreen forest. Small flocks of up to 15 or 20, usually high up in foliage canopy of lofty trees. Commonly suns itself on bare topmost branches early morning and before sunset. Flight swift and powerful but with leisurely-looking wing flaps. Food: exclusively fruit, largely wild figs and nutmegs, swallowed entire. Call: deep, booming, rather mournful *ūk-ook ... ook*, repeated at a few seconds' intervals.

68 SNOW PIGEON *Columba leuconota* Vigors   p. 36
Size Pigeon; 35 cm (14 in.).

Field Characters A particoloured brown and white high-altitude pigeon. In overhead flight blackish head, white body, and white-banded blackish tail are pointers. Seen below observer's level, the blackish head, brown back, white rump-patch, grey wings with three dark bars, and blackish tail with narrow white subterminal band are diagnostic. Sexes alike.

Status, Habitat, etc. Resident at altitudes between 3000 and 4500 m, descending lower in severe winters. Affects rocky cliffs and gorges in the alpine zone and above snow-line. Lives and roosts in colonies on cliffs, small parties flying out to glean on grassy slopes or at edge of melting snows; in summer also in barley fields around upland villages. Flight typical of pigeons, strong, direct, and with rapid wing beats. Food: seeds, grain, bulbils and green vegetable matter. Call: repeated croaks, not unlike hiccups, of recognizable pigeon pattern.

69 HILL PIGEON *Columba rupestris* Pallas
Size Pigeon; length 35 cm (14 in.).

Field Characters A high-altitude pigeon superficially like the familiar Blue Rock but much paler bluish grey above and with whitish under-parts. In flight the pure white rump-patch and broad white band across blackish tail, of pattern like Snow Pigeon's, diagnostic, Sexes alike.

Status, Habitat, etc. Resident in the high Tibetan Plateau facies of extreme . northern Sikkim (probably also Bhutan and Arunachal Pradesh) at 4500 m and above, descending lower in hard winters. Lives in colonies on crags and precipices, commuting to feed in cultivation around upland villages and on spilt grain on traders' caravan routes. Unlike Snow Pigeon, also patronizes crumbling dwellings where available. F o o d : green shoots and grain of barley, oats, etc., and weed seeds. C a l l : high-pitched, quick-repeated *gūt-gūt-gūt-gūt* etc. with a sort of jeering intonation.

70 SPECKLED WOOD PIGEON *Columba hodgsonii* Vigors p. 36
Size Pigeon + ; length 40 cm (16 in.).

Field Characters A dark forest pigeon. M a l e . *Above,* head and neck grey, the latter speckled behind with blackish. Mantle claret-maroon speckled with white on coverts. Rest of upper plumage dark brown and slaty grey; tail blackish brown. *Below,* upper breast speckled with blackish grey and claret, changing to uniform claret on lower breast, passing to deep slaty on belly and under tail-coverts. F e m a l e has the head brownish grey, and red of mantle and breast duller and browner.

Status, Habitat, etc. Resident, subject to nomadic seasonal movements dependent on fruit supply and ripening: tall mixed evergreen forest between 1800 and 4000 m. Small flocks of 6 to 10. Largely frugivorous and arboreal, but also gleans weed seeds and spilt grain in harvested stubbles. C a l l : a very deep *whock-whr-o-o . . . whrroo.*

71 ASHY WOOD PIGEON *Columba pulchricollis* Blyth p. 36
Size Pigeon + ; length 35 cm (14 in.).

Field Characters *Above,* head and nape pale grey; a conspicuous buff collar, broad and black-stippled on hindneck, narrow in front. Upper back and all round lower neck blackish with iridescent green and lilac sheen; lower back and rump blackish leaden grey; wings and tail blackish brown. *Below,* chin white diffusing into buff throat and then into the glossy collar; breast metallic dark slaty blue diffusing to pale cinnamon or brownish buff on belly and vent. Sexes alike.

Status, Habitat, etc. Resident, normally between 1200 and 3200 m; occasionally lower, in the duars: dense moist-deciduous and evergreen forest. Subject to considerable seasonal nomadism as in other fruit pigeons. Pairs, or small flocks of 10 to 30, usually sitting quietly high up in a tree concealed by foliage, hence often overlooked. F o o d : chiefly drupes and berries; also acorns, grain, and seeds. C a l l : a deep sonorous *coo* like that of British Wood Pigeon.

36

Plate 6, artist Paul Barruel

PARAKEETS, CUCKOO, TROGON, FROGMOUTH, PIGEONS

P. BARRUEL

JPIrani 1973

0   3   6   9   12 cm

Plate 7, artist J. P. Irani

SWIFTS, HOUSE MARTINS, OWLS, NUTCRACKER, COUCAL, NIGHTJAR

1 R WHITETHROATED SPINETAIL SWIFT, *Chaetura caudacuta* page 67
  Bulbul +, stouter. Middle back sandy brown. Innermost secondaries partly white. 1250–4000 m.

2 R ALPINE SWIFT, *Apus melba* 68
  Bulbul +, stouter. White underparts; brown breast-band. Up to 2500 m.

3 R HOUSE SWIFT, *Apus affinis* 69
  Sparrow. White rump. Squarish tail. Up to 2000 m.

4 R HOUSE MARTIN, *Delichon urbica* 95
  Sparrow. White underparts. White rump. Forked tail. Up to 5000 m.

5 R NEPAL HOUSE MARTIN, *Delichon nipalensis* 98
  Sparrow ±. Like 4, but throat and vent black. Square tail. 350–4000 m.

6 R SPOTTED SCOPS OWL, *Otus spilocephalus* 56
  Myna. 'Eared'. Spotted crown and nape. Up to 2600 m.

7 R NUTCRACKER, *Nucifraga caryocatactes* 110
  Pigeon ±. Tail white with black wedge in middle. 2000–6000 m.

8 R LESSER COUCAL, *Centropus toulou bengalensis* 55
  House Crow −. Tail-feathers white-tipped. Up to 200 m.

9 R INDIAN SCOPS OWL, *Otus scops* 56
  Myna. 'Eared'. Like Collared Scops but without pale collar on hindneck. Up to 1500 m.

10 R HIMALAYAN JUNGLE NIGHTJAR, *Caprimulgus indicus* 63
  Pigeon −. ♂ A white subterminal spot on 4 pairs outer rectrices. Foothills to 3300 m.

72 **BARTAILED CUCKOO-DOVE** *Macropygia unchall* (Wagler) p. 36
Size Pigeon ± with longer, pointed tail. Overall length 40 cm (16 in.).
Field Characters A slender reddish brown long-tailed forest pigeon.
Male. *Above*, forehead buff, shading to lilac-purple on crown, hindneck
and upper back, highly glossed with metallic green. Rest of upper
plumage rufous, barred with black. Tail rufous, long, broad, graduated,
barred with black. *Below*, chin and throat lilac-buff. Upper breast
metallic lilac diffusing to dull lilac on lower breast and to buff on belly
and vent. Female has upper plumage duller; head and breast barred
with dark brown.
Status, Habitat, etc.   Resident between 450 and 2700 m: dense evergreen
forest and secondary jungle. Pairs or small parties glean in cultivation
clearings and forest glades, or feed up among fruiting trees like green
pigeons. Flight fast, typically pigeon-like, but habit of gliding upward
for settling on a branch, and hind aspect when perched upright, distinctly
cuculine. Food: drupes, berries, grain, weed seeds. Call: normally
silent. A deep, booming *croo-umm* (accent on *umm*) repeated at short
intervals in breeding season.

73 **RUFOUS TURTLE DOVE** *Streptopelia orientalis* (Latham)   p. 33
Size Pigeon: length 35 cm (14 in.).
Field Characters A large dove with rufous-and-blackish scaly patterned
upper plumage, slaty grey rump, a black-and-grey 'chessboard' on
either side of hindneck, and broad grey terminal fringe to blackish
rounded tail—the last particularly conspicuous when spread while
alighting or in display. *Below*, chin and throat whitish; breast pale
vinous-rufous diffusing to white on belly and vent. Sexes alike. Distinguished from the commoner Spotted Dove (74) by rich rufous
coloration and stocky pigeon-like build.
Status, Habitat, etc.   Resident. Duars, foothills and up to 4000 m,
migrating to lower levels and the plains in winter: open mixed forest
and bamboo jungle in broken foothills country, often in the
neighbourhood of cultivation. Ground feeding and granivorous;
gleans in stubbles, earth tracks, etc. Has characteristic aerial display
of the family—springing up in the air with wing-clapping, and coasting
down in an arc or spiral with outspread tail. Food: grain and weed
seeds. Call: a hoarse, mournful sounding *goor* ... *gūr-grūgroo* repeated
in slow runs.

74 **SPOTTED DOVE** *Streptopelia chinensis* (Scopoli)
Size Myna ± ; length 23 cm (9 in.).
Field Characters *Above*, pinkish brown and grey spotted with white; a

conspicuous black-and-white 'chessboard' at base of hindneck. Tail blackish brown and slate, with a broad white terminal fringe conspicuous when spread while alighting. *Below*, vinous grey, paler on throat, white on belly and vent. Sexes alike. Birds from the Mishmi Hills (*edwardi*) are markedly darker.

Status, Habitat, etc. Resident and seasonal local migrant. Duars and normally up to 2400 m, descending to lower elevations and the plains in winter: groves, cultivation and open moist-deciduous jungle. Pairs or small parties glean in stubble fields and on cart tracks etc. Habits and food as in 73 and other granivorous pigeons. Call: a soft, pleasant crooning *krookrūk-krūkroo ... kroo-kroo-kroo*—the final *kroo*s repeated from two to six times.

75 **EMERALD DOVE** *Chalcophaps indica* (Linnaeus)   p. 33
Size Myna + ; length 25 cm (10 in.).
Field Characters A small brownish pink forest dove with brilliant bronzed emerald-green upperparts. Crown and head bluish grey with prominent white forehead and eyebrows, and a band of white-fringed feathers across lower back. Rump grey; tail brown and grey with a broad black band interrupted in the middle. Female differs only in details. Bronze-green mantle, white rump-patch, chestnut underwing and coral red bill conspicuous in a fly-past.

Status, Habitat, etc. Resident (sedentary or local migrant ?). Duars and up to 1800 m in moist-deciduous secondary forest and mixed bamboo jungle, chiefly foothills country. Overgrown *jhoom* clearings, and erosion nullahs through tea gardens are favourite haunts. Singly or pairs, locally not uncommon. Runs about and gleans on jungle tracks etc. Flight silent, swift, direct. Food: grain, weed seeds, berries; occasionally insects, e.g. white ants. Call: a soft, deep moaning note *hoon* repeated at intervals.

PARROTS: Psittacidae

76 **LARGE PARAKEET** *Psittacula eupatria* (Linnaeus)
Size Pigeon ± ; length (including long pointed tail) 50 cm (20 in.); female somewhat smaller.
Field Characters A large grass-green parakeet with typical massive hooked red bill, and a conspicuous deep crimson patch on wing-shoulders (secondary coverts). Male has a prominent rose-pink collar on hindneck, joined to lower mandible by a broad black band. Female lacks both pink collar and black band.

Status, Habitat, etc. Resident; locally nomadic, depending on food supply. Duars and low foothills, rarely as high as 1500 m—forest and wooded country. Small flocks; large congregations where food abundant. Roosts communally in large leafy trees. Flight swift and direct though seemingly unhurried. Food: fruits, vegetables, grains— wild and cultivated. Often damages orchards and grain fields. Call: loud, high-pitched, screaming *keeāk* or *kee-ārr*, deeper and more sonorous than of the commoner Roseringed species; uttered from perch or wing.

77 **REDBREASTED PARAKEET** *Psittacula alexandri* (Linnaeus) p. 36
Size Pigeon − ; overall length (including tail) 40 cm (16 in.).
Field Characters A long and pointed-tailed grass-green parakeet with greyish pink head and wine-red throat and breast. A prominent yellow shoulder-patch where crimson in Large Parakeet (76). Female has the head tinged with blue-green with less plum-coloured bloom; breast redder and darker without vinaceous tinge; bill largely black (*v.* largely yellow in male).
Status, Habitat, etc. Resident, with local nomadic movements governed by food supply. Duars and foothills up to 1500 m—moist-deciduous secondary forest, and neighbourhood of *jhoom* or shifting cultivation. Usually parties of up to 10 or so; occasionally larger rabbles as when raiding ripening paddy or orchard fruit. Flies from branch to branch in a tree with a muffled whirr of wings, and clambers among the fruit-bearing twigs. Feeds quietly, well camouflaged among the foliage; its presence often revealed only by the gentle patter of leaves and gnawed fruit dropping below in the process. Food: fruits, leaf buds, fleshy flower petals, and cereals. Call: short nasal screams, *kaink*, *kaink* ... quickly repeated, often by several birds in chorus as they fly off on disturbance.

78 **EASTERN BLOSSOMHEADED PARAKEET**
Psittacula roseata Biswas
Size Myna − ; overall length (including long pointed tail) 35 cm (14 in.).
Field Characters A slender grass-green parakeet with bright bluish pink and lilac head, a narrow black collar, and red patches on wing-shoulders (coverts). Broad yellowish white tips to narrow blue central tail-feathers conspicuous in flight. Female has the head duller and greyer surrounded by a bright yellow collar. No red shoulder-patches.
Status, Habitat, etc. Resident, with the usual nomadism governed by food supply. Duars and foothills, normally up to 600 m, exceptionally to 1500 m: well-wooded country, light forest and neighbourhood of

jhoom cultivation. Usually small parties; sometimes large congregations at ripening cereal crops. Flight straight, arrowlike, with adroit twists and turns in unison to avoid tree-trunks etc. F o o d : fruits, grains, flower petals, nectar, etc. C a l l : soft musical conversational notes; a shrill *tooi-tooi?* in dashing flight.

79 HIMALAYAN SLATYHEADED PARAKEET
Psittacula himalayana (Lesson)   p. 36

Size Myna; with long pointed tail. Overall length 40 cm (16 in.).

Field Characters A grass-green parakeet, very like 78 but with whole head dark bluish slaty in m a l e. Chin and a narrow black ring encircling neck; a bright verdigris-green collar on hindneck. Prominent dark red patches on wing-shoulders (secondary coverts). Elongated narrow central tail-feathers blue with broad yellow tips, prominent in flight. F e m a l e similar but with paler head and no red shoulder-patches. Both sexes distinguished from 78 by slaty head *v*. bluish pink or grey, and bright yellow tail-tip *v*. whitish.

Status, Habitat, etc. Resident, with marked summer-winter altitudinal movements in addition to the food-finding nomadism. Sikkim and Bhutan; chiefly between 600 and 2500 m, sometimes down to 250 m locally. Replaced in Arunachal Pradesh by an almost identical species (80, q.v.). Keeps to wooded hillsides and valleys in the neighbourhood of terraced cultivation and orchards. Habits and food as in 78. C a l l : flight call *tooi-tooi?* similar to that of 78 but harsher.

80 EASTERN SLATYHEADED PARAKEET
Psittacula finschii (Hume)

Size Myna − . Overall length 40 cm (16 in.).

Field Characters Indistinguishable from 78 in the field. In the hand general coloration paler; back more yellow, less grass-green; central tail-feathers longer, narrower and with duller yellow terminal half.

Status, Habitat, etc. Same as 79, but has a more easterly distribution, from SE. Bhutan through Arunachal Pradesh, Nagaland, Manipur, etc. Habits, food, and calls as in 79, q.v.

81 LORIKEET *Loriculus vernalis* (Sparrman)

Size Sparrow ± ; length 14 cm (5½ in.).

Field Characters A diminutive bright grass-green parrot with crimson rump and short square tail. Sexes alike, but male has a small blue throat-patch which female lacks.

Status, Habitat, etc. Resident and local migrant. Duars and foothills normally up to 1000 m—well-wooded country in evergreen and moist-

deciduous biotope. Also occurs elsewhere in the Peninsula and
Andamans. Pairs or family parties; larger gatherings where food
plentiful. Keeps to the canopy of tall trees, perfectly camouflaged in the
foliage. Clambers energetically among the leaf stalks to feed, often
hanging upside down to rest. Also roosts at night hanging like a bat.
Flight swift and finch-like—several rapid wing strokes followed by a
pause and slight dip. Food: pulp of soft fruits, flower nectar, seeds.
Call: a shrill trisyllabic bat-like squeaking *chi-chi-chee* repeated at
intervals chiefly in flight; also while feeding.

CUCKOOS: Cuculidae

82 REDWINGED CRESTED CUCKOO
Clamator coromandus (Linnaeus)

Size House Crow ± ; slenderer. Length 50 cm (20 in.).

Field Characters *Above*, glossy metallic black with crested head and white
half-collar on hindneck. Wings chestnut as in Crow-Pheasant (97).
Below, throat and breast rusty; rest whitish. Sexes alike. When flying
away from observer, black crest and upper plumage interrupted by
white hind-collar, red wings, and whitish underparts, diagnostic.

Status, Habitat, etc. Resident and partial migrant. Duars and foothills
up to 1500 m. Evergreen and moist-deciduous biotopes, in secondary
forest and scrub-and-bush jungle. Solos. Chiefly arboreal, occasionally
descending to low bushes for food. Very silent except in breeding season.
Flight swift and direct, with quick wing beats like Koel. Brood-parasitic
chiefly on laughing thrushes. Food: mostly caterpillars. Call: harsh
grating screams resembling a jay's; also a whistling double note like
Pigmy Owlet's.

83 LARGE HAWK-CUCKOO *Cuculus sparverioides* Vigors

Size House Crow − ; slimmer. Length 40 cm (16 in.).

Field Characters Appearance at rest and in flight strikingly hawk-like.
Above, ashy grey and ashy brown. Tail greyish brown, banded with
blackish and tipped white or rufous-white. *Below*, throat white, streaked
with ashy and rufous, passing into more rufous upper breast. Rest of
underparts rufous-tinged white, cross-barred with brown. Sexes alike.
Reliably identified only by the distinctive call.

Status, Habitat, etc. Resident; breeds between 900 and 2700 m—wooded
hillsides and valleys. Migrates to peninsular India in winter. Arboreal;
keeps singly to foliage canopy and difficult to see unless it flies. Flight
swift, direct, deceptively shikra-like—a few rapid wing beats followed

by a glide. Hawk-like habit of sweeping upward into branch for alighting adds to the deception. Obstreperously vocal in breeding season (summer). Brood-parasitic chiefly on laughing thrushes. Food: caterpillars, beetles and other insects. Call: a loud, shrill, screaming crescendo whistle *pipeeah* (accent on 2nd syllable) in runs of 3 to 6; repeated with monotonous persistency all day and during night.

84 HAWK-CUCKOO or BRAINFEVER BIRD *Cuculus varius* Vahl

Size Pigeon $\pm$; slenderer, with proportionately longer tail. Length 35 cm (14 in.).

Field Characters A shikra-like cuckoo, in general very similar to 83. *Above*, ashy grey. Tail tipped rufescent with 4 or 5 whitish and black bands, the terminal one broadest. *Below*, white, tinged with rufous and ashy on breast; barred with brown on belly and flanks. Sexes alike.

Status, Habitat, etc. Same as 83, but inhabits a lower altitudinal zone— normally below 1000 m. Affects deciduous and semi-evergreen biotope— secondary forest, groves near habitations and cultivation, etc. Habits not different from 83. Flight and movements likewise deceptively shikra-like. Silent in winter, thus apt to be overlooked; very noisy in breeding season (summer). Brood-parasitic mainly on jungle babblers and laughing thrushes. Food: caterpillars and other insects. Call: a loud, shrieking, high-pitched whistle *wee-piwhit*, well syllabified as *brainfever* (accent on *fe*), in runs of 4 to 6 crescendo calls repeated *ad nauseam* all day and during night.

85 HODGSON'S HAWK-CUCKOO *Cuculus fugax nisicolor* Blyth

Size Pigeon − ; slenderer. Length 30 cm (12 in.).

Field Characters Differs from Brainfever Bird (84) chiefly by its unbarred rufous underparts. *Above* (including sides of head), slaty grey. Wings brownish; tail alternately banded grey and black, tipped rufous. *Below*, chin grey; throat and foreneck white, sparsely streaked with grey. Rest of underparts largely rufous; under tail-coverts white. Sexes alike.

Status, Habitat, etc. Resident between 600 and 1800 m, subject to considerable altitudinal and local migration. Affects wooded hillsides and valleys in deciduous, semi-evergreen and evergreen biotopes, keeping chiefly to the understorey. Brood-parasitic, but other habits little known. Usually silent (in winter) and skulking, thus doubtless often overlooked. Food: caterpillars, cicadas, and other insects. Call: a sibilant, insistent *gee-whizz* repeated up to 20 times, reminiscent of the high-pitched shrieking of Large Hawk-Cuckoo (83) but still more shrill. Bird noisy in breeding season.

44

Plate 8, artist J. P. Irani

BARBETS, BEE-EATERS, ROLLERS, KINGFISHERS, TROGON

1 R   LINEATED BARBET, *Megalaima lineata*   page 78
    Myna + . Pale streaked head and neck. Naked yellow
    patch round eye. Up to 800 m.

2 R   BLUETHROATED BARBET, *Megalaima asiatica*   79
    Myna. Black transverse band on crimson crown. Up
    to 2000 m.

3 R   CHESTNUTHEADED BEE-EATER, *Merops*
    leschenaulti   73
    Bulbul ± . Yellow chin and throat. Up to 1500 m.

4 R   ROLLER, *Coracias benghalensis* (*affinis*)   74
    Pigeon. Throat blue-streaked. Breast purplish. Up to
    600 m.

5 R   BLUEBEARDED BEE-EATER, *Nyctyornis athertoni*   74
    Pigeon — . Tail square-cut; no pins. Up to 1700 m.

6 R   BROADBILLED ROLLER, *Eurystomus orientalis*   75
    Pigeon. Very broad orange-red bill and legs. Up to
    1000 m.

7 R   HIMALAYAN PIED KINGFISHER, *Ceryle lugubris*   70
    House Crow ± . Prominent crest. Black and rufous-
    brown breast-band. Up to 2000 m.

8 R   INDIAN RUDDY KINGFISHER, *Halcyon coromanda*   72
    Myna + . White rump-patch (in flight). Bright red bill
    and feet. Up to 1800 m.

9 R   WARD'S TROGON, *Harpactes wardi*   70
    Pigeon ± . Belly in male crimson-pink; in female
    primrose-yellow. Graduated rectrices squarely
    truncated. 1500–3000 m.

1

3

4

5

6

9

♀

7

8

JPIrani 1972

0   3   6   9   12   15 cm

Winston Creado

Plate 9, artist Winston Creado

HORNBILLS, WOODPECKERS, PITTA

1 R **RUFOUSNECKED HORNBILL,** *Aceros nipalensis*   page 76
Vulture. Up to 1800 m.

2 R **WREATHED HORNBILL,** *Rhyticeros undulatus*   76
Vulture ±. Up to 2400 m.

3 R **GREAT PIED HORNBILL,** *Buceros bicornis*   77
Vulture. Concave-topped casque. Black breast. White
tail with single black band. Up to 2000 m.

4 R **PALEHEADED WOODPECKER,** *Gecinulus grantia*   87
Myna +. ♂ Crimson-pink patch on crown. ♀ without
crown-patch. Up to 1000 m.

5 R **LARGER GOLDENBACKED WOODPECKER,**
Chrysocolaptes lucidus   90
Pigeon. Distinctive black-and-white patterned throat.
Crimson rump. ♂ Crimson crest. ♀ White-stippled
black crest. Up to 1600 m.

6 R **BLACKNAPED GREEN WOODPECKER,** *Picus canus*   85
Pigeon ±. ♀ Forehead and forecrown black. Nape
grey-streaked. Up to 2100 m.

7 R **CRIMSONBREASTED PIED WOODPECKER,**
Picoides cathpharius   88
Bulbul −. ♀ Crown entirely black. 1700–4000 m.

8 R **HOODED PITTA,** *Pitta sordida*   92
Quail. Up to 2000 m.

86 INDIAN CUCKOO *Cuculus micropterus* Gould

Size Pigeon $\pm$; slenderer, with proportionately longer tail. Overall length 35 cm (14 in.).

Field Characters Very similar to Cuckoo (87), q.v. *Above*, dark slaty with a brownish tinge. *Below*, pale ashy and white, cross-barred with widely spaced black bands. Readily distinguished from *C. canorus* (87) by broad black subterminal band on tail; conclusively by the unmistakable call (see below). Sexes alike.

Status, Habitat, etc. Summer visitor to the duars and hills normally up to 2300 m, occasionally to 3500 m : openly wooded country. Very noisy between April and August, silent in other months when liable to be overlooked. Mainly arboreal, keeping singly to foliage canopy, flying about hawk-like above the tree-tops. Particularly obstreperous in early morning and at dusk—often calling throughout moonlit nights. Also calls on the wing in courtship chase. Brood-parasitic mainly on drongos. Food : caterpillars and other insects, sometimes picked off the ground while hopping awkwardly. Call : distinctive and diagnostic, a loud fluty 4-syllabled whistle—variously syllabified as *Crossword-puzzle*, *Orange-pekoe*, *Bo-kotāko*, *Kyphăl-păkka*, etc.—repeated intermittently in monotonous runs of several minutes, for hours on end.

87 CUCKOO *Cuculus canorus* Linnaeus

Size Pigeon $\pm$; slenderer. Length 35 cm (14 in.).

Field Characters Superficially, in the long pointed wings and speed and style of flight, very hawk-like. Male. *Above*, dark ashy grey. Tail blackish brown, spotted and tipped with white; no subterminal black band as in Indian Cuckoo (86). *Below*, chin, throat and breast pale ashy; rest white, narrowly cross-barred with blackish. Female has rufous tinge on throat and breast. Occasionally found in a hepatic phase with entire upperparts, throat and upper breast barred chestnut-and-black; lower breast and belly rufous-tinged.

Status, Habitat, etc. Resident, nomadic, and locally migratory. Affects openly wooded country, hill orchards, etc., breeding at between 600 and 4000 m. Brood-parasitic on a wide range of hosts—pipits, shrikes, babblers, flycatchers, chats. Habits similar to Indian Cuckoo (86) and others of the family. Silent except in spring and summer when breeding; then very noisy. Food : caterpillars, cicadas and other insects. Call : the well known *cūck-koo*, occasionally varied by *cūck-cūck-koo*, repeated persistently at about one per second in long unbroken runs. Also has some hoarse wheezy chuckles. Female gives a series of 'water-bubbling' notes *quick-quick*, *quick-quick*, etc., indistinguishable from those of female *C. micropterus* (86) and *C. saturatus* (88).

88 HIMALAYAN CUCKOO *Cuculus saturatus* Blyth

Size Pigeon − ; slenderer. Length 30 cm (12 in.).

Field Characters Indistinguishable from the common Cuckoo (87) q.v., except by its distinctive 4-noted call (see below). Females sometimes also hepatic as in 87.

Status, Habitat, etc. Resident or summer visitor?. Uncertainty due to its being silent between about August and March when likely to be overlooked or confused with 87. Affects hilly wooded country and orchards, etc., breeding at between 1500 and 3300 m. Brood-parasitic mainly on leaf- and flycatcher-warblers (*Phylloscopus* and *Seicercus*). Habits and food as in 87. Call: distinctive and entirely diagnostic— a loud, far-carrying *oop-poop-poop-poop* (accent on initial *oop*) preceded by a soft undertone *ūp*, only audible at close range. Easily confused with call of hoopoe, but is characteristically 4-noted against the hoopoe's three. Less obstreperous than both Indian Cuckoo (86) and Cuckoo (87).

89 SMALL CUCKOO *Cuculus poliocephalus* Latham

Size Myna + ; with longer tail. Length 25 cm (10 in.).

Field Characters A smaller edition of the common Cuckoo (87). *Above*, slaty grey. *Below*, buffy white, cross-barred with black. Sexes nearly alike, but female also has a hepatic phase. Most reliably identified by the husky call (see below).

Status, Habitat, etc. Summer visitor and/or resident?; locally common in well-wooded country. Breeds at between 1500 and 3200 m, possibly higher. Brood-parasitic on small ground-nesting passerines such as leaf warblers (*Phylloscopus*), wren-babblers (*Pnoepyga*) and shortwings (*Brachypteryx*). Mostly silent between August and April, thus liable to be overlooked. Habits and food typical of the family, cf. 86, 87. Very noisy in breeding season (May-July), calling persistently throughout the day, particularly if cloudy overcast, and at night; from a perch as well as on the wing. Call: a curious husky chattering of 5 or 6 unmusical notes, the first half rising in scale the second falling. Well-syllabified as *That's your choky pepper ... choky pepper* (accent on first *choky*), quickly repeated.

90 BAYBANDED CUCKOO *Cacomantis sonneratii* (Latham)

Size Myna ± ; slenderer. Length 24 cm (9½ in.).

Field Characters A small slim cuckoo. *Above*, bright rufous or bay, conspicuously cross-barred with brown. Tail largely rufous, the feathers tipped white and subtipped black. *Below* (including sides of head and neck) whitish, with fine wavy brown cross-bars. Sexes alike. Easily confused with hepatic female of Plaintive Cuckoo (91), but calls usually diagnostic.

Plate 10, artist Paul Barruel

WOODPECKERS, PICULET, BARBETS, HONEYGUIDE

1 R GOLDENBACKED THREETOED WOODPECKER,
Dinopium shorii   page 86
Pigeon ±. Crimson rump. Black tail. White band
down each side of neck. ♀ similar but crown and crest
white-streaked black. Duars to 700 m.

2 R DARJEELING PIED WOODPECKER, *Picoides*
darjellensis   88
Myna ±. ♀ Hindcrown and nape black. 1700–4000 m.

3 R REDEARED BAY WOODPECKER, *Blythipicus*
pyrrhotis   90
Pigeon −. ♀ Without scarlet nape-band. Duars to
2000 m.

4 R RUFOUS PICULET, *Sasia ochracea*   84
Sparrow −. Stub-tailed. White supercilium. ♀
Without golden forehead. Duars to 2100 m.

5 R GREAT SLATY WOODPECKER, *Mulleripicus*
pulverulentus   87
House Crow +. ♀ Without crimson moustachial
stripe. Duars to 2000 m.

6 R LARGE YELLOWNAPED WOODPECKER, *Picus*
flavinucha   85
Pigeon ±. ♀ Chin and throat rufous-brown. Up to
2400 m.

7 R GOLDENTHROATED BARBET, *Megalaima franklinii*   79
Myna. 600–2400 m.

8 R GREAT BARBET, *Megalaima virens*   78
Myna +. 1000–3000 m.

9 R ORANGERUMPED HONEYGUIDE, *Indicator*
xanthonotus   82
Sparrow. ♀ Duller. 1500–3500 m.

Inches

0 1 2 3 4

P. BARRUEL

F. BARRUEL

Inches

Plate 11, artist Paul Barruel

SUNBIRDS, FLOWERPECKER, BROADBILLS, PITTA

1 R **MRS GOULD'S SUNBIRD,** *Aethopyga gouldiae* page 233
Sparrow – . Sides of head crimson. Elongated central
rectrices metallic purple-blue. ♀ Crown grey. Rump
yellow. Belly yellow; throat grey. Tail short. Duars to
3300 m.

2 R **BLACKBREASTED SUNBIRD,** *Aethopyga saturata* 233
Sparrow – . Head dark. Elongated central rectrices
metallic purple. ♀ Olive above: grey crown, yellow
rump; short tail. 450–2000 m.

3 R **FIRETAILED YELLOWBACKED SUNBIRD,**
Aethopyga ignicauda 234
Sparrow – . Back and tail, with elongated central
rectrices, bright red. ♀ Olive. No yellow band across
rump. Tail short. 1200–4000 m.

4 R **FIREBREASTED FLOWERPECKER,** *Dicaeum*
ignipectus 232
Sparrow – . ♀ Olive-green; rump yellower. Buff below,
tinged olive on sides. 750–3000 m.

5 R **LONGTAILED BROADBILL,** *Psarisomus dalhousiae* 91
Bulbul + . Duars to 2000 m.

6 R **COLLARED BROADBILL,** *Serilophus lunatus* 91
Bulbul ± . ♀ Similar, but with a broken whitish collar
on sides of neck. Duars to 1700 m.

7 R **BLUENAPED PITTA,** *Pitta nipalensis* 92
Quail + . ♀ Similar, but hindcrown rufous; hindneck
green. Duars to 2000 m.

Status, Habitat, etc.   Resident, nomadic or seasonal migrant? Uncertainty due to its silence in non-breeding season and possibility of being over-looked then. Duars and foothills, occasionally as high up as 2400 m: lightly wooded and cultivated country as well as heavy forest, in moist-deciduous and evergreen biotope. Arboreal; insectivorous. Keeps singly to bare tree tops when calling. Flight, general habits, and food as of the family. Brood-parasitic on bulbuls and small babblers. Call: a loud, pleasant 4-noted whistle *weeti-teeti* repeated with monotonous persistency, tail depressed, wings drooped at sides; somewhat reminiscent of *crossword-puzzle* call of Indian Cuckoo. Song, a sweet clear whistling *tee titee-teeti titee-teeti?*

91 PLAINTIVE CUCKOO *Cacomantis merulinus* (Scopoli)
Size Myna ± ; slimmer. Length 23 cm (9 in.).

Field Characters A small, slim cuckoo. *Above*, slaty and brown. Tail blackish, with white tips to the feathers, the outer ones also barred with white. *Below*, chin, throat and breast grey. A white patch on underwing (at base of primaries), conspicuous in flight. Sexes alike but female also has a hepatic phase, then easily confusable with Baybanded (90), q.v.

Status, Habitat, etc.   As in 90, silent in non-breeding season (June to September) whence uncertain whether resident, nomadic or visitor. Duars and foothills, occasionally up to 2700 m: light open forest, village groves, tea gardens, etc. Keeps singly to foliage canopy, mounting to exposed topmost branches for calling. Active and restless, constantly flying from tree to tree to call from different vantage points. Brood-parasitic on wren-warblers, tailor birds, etc. Food: caterpillars, bugs, and other insects. Call diagnostic: a plaintive whistle *piteer* or *kiveer* repeated at short intervals. Another call of *crossword-puzzle* pattern but in higher key—*weeti-teeti* or *peter-peter*. Song, a clear lilting whistling of several notes ending interrogatively: *pee .. pipeepee-pipeepee?*. Calls particularly in overcast weather; also during night.

92 EMERALD CUCKOO *Chalcites maculatus* (Gmelin)   p. 36
Size Sparrow; length 18 cm (7 in.).

Field Characters A diminutive resplendent cuckoo. Male. *Above*, brilliant glossy bronze-green. A white patch on wing conspicuous in flight. Tail-feathers tipped white, the outermost pair with three white bars. *Below*, chin, throat and upper breast like back; rest white, barred with metallic bronze-green. Under tail-coverts metallic green barred with white. Female. *Above*, glistening emerald green with golden rufous crown and nape. Tail barred chestnut and black, tipped with white. *Below*, white, tinged with rufous on throat and flanks; barred with bronze-brown narrowly on throat, broadly on belly.

Status, Habitat, etc. Status uncertain. Possibly resident but subject to nomadism and/or seasonal migration like other cuckoos. Duars and foothills, normally up to 1000 m. Keeps singly or in small parties of 3 or 4 to the foliage canopy in secondary evergreen jungle; its presence usually detected by its calls or when it makes occasional aerial sallies after winged insects. Brood-parasitic on sunbirds and spiderhunters. Food: caterpillars, bugs and other insects, and spiders. Call: a clear, high-pitched whistled trill or rattle of 3–6 notes rapidly uttered, reminiscent of Lorikeet.

93 **DRONGO-CUCKOO** *Surniculus lugubris* (Horsfield)   p. 53

Size Myna ± ; slimmer, with long (forked) tail. Length 25 cm (10 in.).

Field Characters Glossy metallic black. General appearance deceptively like Black Drongo, but under tail-coverts and base of outermost rectrices nearly always barred with white. Sexes alike. Calls diagnostic. (See below.)

Status, Habitat, etc. Resident. Also nomadic and/or locally migratory. Movements poorly known owing to absence of calling during non-breeding season. Duars and foothills up to 2000 m, in well-wooded country. Arboreal. Keeps singly to foliage canopy of trees, mounting to exposed top branches for calling. Flight cuckoo-like, noticeably different from drongo's. Sometimes catches winged insects by springing up into air like drongo. Brood-parasitic reportedly on drongos, minivets, etc., but biology little known. Food: caterpillars, grasshoppers, wild figs, etc. Call: song quite distinctive—a run of 5 or 6 evenly spaced whistles *pip-pip-pip-pip-pip-pip* (as if the bird was counting 1-2-3-4-5-6 or practising the musical scale) rising in pitch with each successive *pip* and breaking off abruptly. Reiterated monotonously after a few seconds.

94 **KOEL** *Eudynamys scolopacea* (Linnaeus)

Size House Crow; slimmer, with longer tail. Length 45 cm (18 in.).

Field Characters M a l e, overall glistening metallic black, with yellowish green bill and bright crimson eyes. Call diagnostic (see below). F e m a l e, dark brown spotted, barred and mottled with white.

Status, Habitat, etc. Resident and marked local migrant. Duars and foothills, occasionally to 1800 m: lightly wooded country and groves around villages and cultivation—usually wherever House and Jungle Crows occur. Arboreal. Keeps singly to leafy trees. Very vocal in summer; largely silent in winter hence often overlooked. Brood-parasitic chiefly on crows. Food: fruits and berries; insects and other small living creatures. Call: male utters loud shrill shrieking whistles *kūoo, kūoo*, etc. rising with each repetition to frantic pitch, and breaking

52

Plate 12, artist K. P. Jadav

SWALLOWS, SHRIKES, ORIOLES, TREE PIE, SKYLARK

1 R **REDRUMPED SWALLOW,** *Hirundo daurica*   page 95
      Sparrow ±. Chestnut collar on hindneck. Rump pale
      chestnut. Streaked underparts. Up to 2500 m.

2 M **TYTLER'S SWALLOW,** *Hirundo rustica tytleri*   94
      Sparrow ±. Chestnut underparts. Pectoral collar
      rudimentary. Up to 3000 m.

3 R **ASHY SWALLOW-SHRIKE,** *Artamus fuscus*   103
      Bulbul ±. Bluish finch-like bill. Slaty black white-
      tipped tail. Up to 1700 m.

4 M **BLACKNAPED ORIOLE,** *Oriolus chinensis*   100
      Myna. Up to 2000 m.

5 R **HIMALAYAN TREE PIE,** *Dendrocitta formosae*   109
      Myna ±. White wing-patch. Chestnut vent. Up to
      2300 m.

6 M **BROWN SHRIKE,** *Lanius cristatus*   99
      Bulbul ±. White supercilium. Black eye-stripe. No
      wing-mirror. Up to 1800 m.

7 R **BLACKHEADED SHRIKE,** *Lanius schach tricolor*   98
      Bulbul +. White wing-mirror. Tail tipped and
      margined rufous. Up to 4000 m.

8 M **SMALL SKYLARK,** *Alauda gulgula*   94
      Sparrow. Short tuft-like crest. Above 1600 m.

9 R **GOLDEN ORIOLE,** *Oriolus oriolus*   99
      Myna. Narrow black streak through eye. Duars.

0 3   6   9   12   15 cm

K. P. Jadav

0   3   6   9   12   15   18 cm

JPIrani 1973

Plate 13, artist J. P. Irani

DRONGOS, THRUSH, MAGPIE-ROBIN, MUNIA, WREN

1 R BRONZED DRONGO, *Dicrurus aeneus* page 102
Bulbul +. Highly glossed green and blue. Up to 2000 m.

2 R LARGE BROWN THRUSH, *Zoothera monticola* 210
Myna +. Very large curved bill. Prominent buff under-wing patch (in flight). Foothills to 3000 m.

3 R BLACK DRONGO, *Dicrurus adsimilis* 101
Bulbul +. Deeply forked tail. Thinly wooded and agricultural habitat. Up to 2000 m.

4 R DRONGO-CUCKOO, *Surniculus lugubris* 51
Myna ±. Very like 3, but vent white-barred. Bill slenderer; nostrils exposed, circular. Up to 2000 m.

5 R GREY DRONGO, *Dicrurus leucophaeus* 101
Bulbul +. Confusable with 3, but slaty black with unglossed underparts. Wooded and forest habitat. Up to 3000 m.

6 R HAIRCRESTED DRONGO, *Dicrurus hottentottus* 103
Myna ±. Outermost rectrices upcurled. Up to 2000 m.

7 R LESSER RACKET-TAILED DRONGO, *Dicrurus remifer* 102
Myna +. Wirelike outermost rectrices tipped with flat spatulae. Bushy frontal tuft. Up to 2000 m.

8 R LARGE RACKET-TAILED DRONGO, *Dicrurus paradiseus* 103
Myna. Wirelike outermost rectrices tipped with curled spatulae. Full backward-curving frontal crest. Up to 1500 m.

9 R MAGPIE-ROBIN, *Copsychus saularis* 196
Bulbul. Graduated black-and-white cocked tail. Up to 1900 m.

10 R WHITEBACKED MUNIA, *Lonchura striata* 238
Sparrow −. Pale-streaked back. White rump. Pointed tail. Up to 1800 m.

11 R WREN, *Troglodytes troglodytes* 214
Sparrow −. Diminutive. Narrowly barred plumage. Short erect tail, 2200–4700 m.

off abruptly at the 7th or 8th. This song repeated monotonously over and over again.

95 LARGE GREENBILIED MALKOHA *Rhopodytes tristis* (Lesson)

Size Pigeon ± ; with tail 40 cm long. Overall length 50 cm (20 in.).

Field Characters A long-tailed ashy grey and green-glossed cuckoo with naked crimson patches round eyes and yellowish green bill. Green-glossed black graduated tail with broad white tips to the feathers very conspicuous, especially in flight and when spread before alighting. Sexes alike.

Status, Habitat, etc.   Resident. Duars and foothills, occasionally up to 1800 m: dense thickets and shrubbery in tropical evergreen jungle. Keeps singly. Rather sluggish, skulking, and addicted to dense thickets. Creeps and weaves its way through the tangles with remarkable celerity in search of food, occasionally taking short laboured flights from one thicket to another. A non-parasitic cuckoo. Builds a crow-like nest of twigs and rears its own brood. F o o d : caterpillars and large insects; lizards and other small living creatures. C a l l : peculiar low croaks *ko, ko, ko*, etc. uttered at short intervals; a low chuckle when flushed.

96 SIRKEER CUCKOO *Taccocua leschenaultii* Lesson

Size House Crow ± ; with longer, broader graduated tail. Overall length 45 cm (18 in.).

Field Characters A heavy-tailed earthy brown and rufous cuckoo (reminiscent of crow-pheasant) with fine glistening black shaft-streaks to head and breast feathers, and stout curved cherry-red and yellow bill. Broad white tips to the blackish cross-rayed tail, conspicuous in flight. Sexes alike.

Status, Habitat, etc.   Resident. Duars, foothills, and up to 2000 m: deciduous sparsely scrubbed broken country with ravines and thorn-and-grass jungle. Largely terrestrial. Keeps singly or in pairs, threading its way through thickets like a crow-pheasant. Also ascends trees in search of food, hopping from branch to branch with great agility. Runs swiftly through undergrowth like a mongoose when disturbed. Flight feeble, usually short and seldom undertaken. A non-parasitic cuckoo, building a crow-like nest of twigs and rearing its own brood. F o o d : insects, lizards and other small animals; also fallen berries. C a l l : occasional; sharp, loud *kek-kek-kek-kerek-kerek-kerek* of quality of a parakeet's shrieks.

97 CROW-PHEASANT or COUCAL *Centropus sinensis* (Stephens)

Size Jungle Crow ± with long, broad graduated tail. Overall length 50 cm (20 in.).

Field Characters A stout, glossy black ground cuckoo with conspicuous chestnut wings and broad, graduated black tail. Sexes alike.

Status, Habitat, etc.   Resident. Duars, foothills, and locally up to 2000 m: deciduous scrub and secondary jungle, tall grassland, tea gardens, orchards, etc. Largely terrestrial; solitary or pairs. Stalks about through undergrowth like a pheasant, methodically searching for food, or hops its way up from branch to branch of trees and shrubs in the quest. Is poor on the wing, usually flying only short distances from one patch of shrubbery to the next. Non-parasitic; building a globular twig nest and rearing its own brood. Food: caterpillars, large insects, lizards, young mice, etc. Particularly destructive to small birds' eggs and nestlings. Call: a deep, resonant, quick-repeated *coop-coop-coop-coop* etc. in runs of 6 to 7 or more, sometimes given as an uneven duet by two widely spaced individuals. Also a variety of harsh croaks and gurgling chuckles.

98 **LESSER COUCAL** *Centropus toulou bengalensis* (Gmelin)   p. 37
Size House Crow – ; with long, broad, graduated tail. Overall length 35–40 cm (14–16 in.).

Field Characters Similar to Coucal (97)—glistening blue-black with chestnut wings—but appreciably smaller and with the tail-feathers white-tipped. Sexes alike, but female larger.

Status, Habitat, etc.   Resident (possibly partially migratory). Duars and submontane tracts, normally below 200 m. Confined to expanses of tall grassland mixed with scrub, interspersed in forest; thus absent over large tracts of country. Habits not markedly different from Coucal. A non-parasitic terrestrial cuckoo. Flight laboured and ill-sustained— several rapid wing beats followed by a glide. Food: mainly grasshoppers and lizards. Call: a double series of rather ventriloquistic notes *whoot*, *whoot*, *whoot*, *whoot* followed after a short pause by *kurook*, *kurook*, *kurook*, *kurook*.

OWLS: Strigidae

99 **BAY OWL** *Phodilus badius* (Horsfield)   p. 16
Size Myna + ; length 30 cm (12 in.).

Field Characters A small dainty chestnut-bay owl (spotted above with black and buff), with short ear-like tufts projecting above sides of head, and fully feathered legs. Vinous pink facial disc and surrounding white ruff stippled with chestnut and black. *Below*, vinous pink, largely spotted with black and white. Sexes alike.

Status, Habitat, etc.   Rare resident. Duars and foothills, locally up to

1500 m. Confined to heavy evergreen forest. Strictly nocturnal and habits very little known. Hides in dark tree-holes in daytime; flies and hunts in complete darkness. Food: small mammals, small birds, lizards, frogs, large insects, etc. Call: described as 'a loud 3-noted whistle reminiscent of someone calling his dog'.

100 **SPOTTED SCOPS OWL** *Otus spilocephalus* (Blyth) p. 37
Size Myna; length 18–20 cm (7–8 in.).
Field Characters A small rusty brown 'eared' forest owlet, seldom seen; recognized mainly by its distinctive calls, q.v. *Above*, crown and nape with conspicuous twin spots of black and white. Face whity brown, indistinctly barred, encircled by a buffish ruff. Wing- and tail-quills brown profusely pale-banded. *Below*, speckled and stippled brown and white. Sexes alike.
Status, Habitat, etc. Resident. Duars, foothills, and up to 2600 m (subspecies *spilocephalus*). In dense submontane evergreen jungle; oak, rhododendron, and pine forest higher up. Habits little known since bird entirely nocturnal. Hides during daytime in tree-holes, and heard only after dusk. Particularly vocal during breeding season, March to June. Food: large insects, lizards, etc.; possibly also small rodents and small birds. Call: a metallic cowbell-like double whistle *phew-phew*, or *tunk-tunk*, repeated at a few seconds' intervals throughout the night, several birds within earshot answering one another.

101 **INDIAN SCOPS OWL** *Otus scops* (Linnaeus) p. 37
Size Myna; length 19 cm (7½ in.).
Field Characters A small slim 'eared' owl, delicately vermiculated greyish brown, with fully feathered legs and prominent ear-tufts. Difficulty of identifying in the field enhanced by occurrence of dark and light, and rufous and grey colour phases. Distinguished from similar Collared Scops (102) by absence of pale collar at base of hindneck.
Status, Habitat, etc. Resident. Submontane tracts and duars, and locally up to 1500 m (subspecies *sunia*). Affects deciduous and evergreen forest, orchards, groves of trees around villages and cultivation, etc. Entirely nocturnal, retiring during daytime into hollows in tree-trunks or densely foliaged branches, standing upright and sleeked with half-shut eyes. Emerges to hunt at dusk and retires into seclusion again before dawn, thus its biology little known. Food: large insects, lizards, mice, small birds. Call: variously described: the song as a monotonous *kūrook-took*, or *wūkh-tuk-ta*, or *wūk-chug-chug*, intermittently repeated all through the night.

102 **COLLARED SCOPS OWL** *Otus bakkamoena* Pennant p. 16
Size Myna ± ; length 23–25 cm (9–10 in.).
Field Characters A small 'eared' owl, very similar to Indian Scops (101); differentiated from it chiefly by presence of a prominent pale collar on hindneck (upper back). *Above*, grey-brown or rufous-brown, mottled and vermiculated with whitish. *Below*, chin and throat whitish, the latter barred and stippled with black. Rest of underparts white to rich buff, streaked with black and with fine wavy bars of reddish brown. Sexes alike.
Status, Habitat, etc. Resident. Duars, foothills, and up to 2400 m (subspecies *lettia*)—open forest of sal, oak and pine; also groves of trees and bamboo around habitations and cultivation. Entirely nocturnal, thus oftener heard than seen. Food: large insects, lizards, mice. Call: a soft, interrogative *wūt?* or *what?* repeated monotonously every few seconds over long periods between dusk and dawn.

103 **FOREST EAGLE-OWL** *Bubo nipalensis* Hodgson p. 16
Size Kite ± ; stouter built. Length 60 cm (24 in.).
Field Characters A large, powerful nocturnal owl with two projecting black-and-white ear-tufts, fully feathered legs, and brown eyes. *Above*, dark brown, scalloped with buff. *Below*, fulvous-white barred with blackish on throat and breast, and chevron-like marks on belly. Sexes alike.
Status, Habitat, etc. Resident. Submontane tracts, duars and foothills normally up to 900–1200 m, locally to 2000 m: tropical moist-deciduous or evergreen forest. Solos or pairs. Largely nocturnal, spending the daytime on some leafy bough in deep forest. A bold and powerful hunter, pouncing on large game birds such as peafowl and junglefowl at their nightly roosts. Food: also takes small mammals—hares, newly born fawns, etc.; lizards, snakes and fish. Call: a deep, eerie, moaning hoot. The long-drawn kite-like whistle circumstantially attributed to it needs verification.

104 **TAWNY FISH OWL** *Bubo flavipes* (Hodgson)
Size Kite ± ; stouter built. Length 60 cm (24 in.).
Field Characters A large orange-rufous 'eared' owl, heavily streaked with blackish on upperparts. Much buff on scapulars and wing-coverts. *Below*, rich orange-rufous with dark brown shaft-stripes, broadest on breast. A white throat-patch. Legs unfeathered; eyes yellow. Sexes alike.
Status, Habitat, etc. Resident. Submontane tracts, duars and foothills up to 1500 m: outscoured forested banks of hill streams especially where

they debouch into the plains. Pairs. Crepuscular and partially diurnal, sometimes even hunting in daytime. Perches in leafy trees overlooking water. Swoops down to capture fish from near surface. Food: mainly fish. Also crabs, rodents, lizards, and large insects. Call: a deep *whoo-hoo*; also a cat-like mewing note.

105 COLLARED PIGMY OWLET *Glaucidium brodiei* (Burton) p. 16

Size Myna − ; length 17 cm (6½ in.).

Field Characters A diminutive owl without ear-tufts; barred grey-brown overall, with prominent white supercilium and throat-patch. A rufous half-collar at base of hindneck. Sexes alike. Also has a dimorphic colour phase, rufous or chestnut. From behind, the half-collar, with a black spot on each side of nape, looks deceptively like a staring owl's face!

Status, Habitat, etc. Resident. Submontane tracts, duars, foothills, and up to 3200 m: open forest of oak, rhododendron, fir, deodar, etc. Usually solo. Crepuscular and markedly diurnal, flying about freely and even hunting in daytime. Bold and fierce for its size, occasionally capturing birds almost as large as itself. Food: small birds, mice, lizards, and large insects. Call: a pleasant 4-noted bell-like whistle *toot ... tootoot ... toot* in runs of 3 or 4, repeated intermittently all hours of day. The call often ends with only the *tootoot*.

106 HIMALAYAN BARRED OWLET *Glaucidium cuculoides* (Vigors)

Size Myna ± ; length 23 cm (9 in.).

Field Characters A dumpy, 'hornless' dark olive-brown owlet, closely barred with whitish above and below. Belly whitish, with fine dark striations; a prominent white throat-patch. Sexes alike.

Status, Habitat, etc. Resident. Duars, foothills, and up to 2700 m: tropical and subtropical evergreen jungle at low elevations; open oak, rhododendron and fir forest higher up (chiefly subspecies *austerum*). Solos. Largely diurnal, moving about freely and hunting in broad daylight. Invariably subjected to mobbing by small birds. Flight bounding and dipping (as of other owlets)—a few rapid wing beats followed by a pause with wings closed. Food: small birds, mice, lizards, large insects. Call: a crescendo of harsh squawking. Song (breeding) a prolonged bubbling musical whistle.

107 BROWN HAWK-OWL *Ninox scutulata* (Raffles)

Size Pigeon ± ; length 30 cm (12 in.).

Field Characters A very hawk-like owl. *Above*, dark greyish brown with whitish forehead and irregular white patches about the shoulders.

Below, throat and foreneck fulvous, streaked with brown; rest of underparts white with large reddish brown drops forming broken bars. Tail barred with black, tipped with white. Sexes alike.

Status, Habitat, etc. Resident. Duars, foothills, and up to 1000 m: forest lining streams, and well-wooded country and groves often near habitations. Solo or pairs. Crepuscular and nocturnal. Spends daytime in seclusion of dense leafy trees, but if disturbed flies out in bright sunlight without apparent discomfort. Flight and manner of sweeping upward to alight on a branch strikingly hawk-like. Often captures winged insects in the air. Food: large insects, small birds and bats, mice, lizards, frogs. Call: a distinctive soft musical *oo ... ūk, oo ... ūk, oo ... ūk*, etc. in runs of 6 to 20, repeated with a short pause between each run. Very vocal in breeding season, particularly during moonlit nights.

108 **SPOTTED OWLET** *Athene brama* (Temminck)

Size Myna $\pm$; length 21 cm ($8\frac{1}{2}$ in.).

Field Characters A common white-spotted greyish brown owlet with typical large round head (no ear-tufts) and forwardly directed staring yellow eyes. Sexes alike.

Status, Habitat, etc. Resident (subspecies *indica* and *ultra*). Submontane tracts, duars, foothills and locally up to 1400 m: groves of ancient trees, deserted buildings, etc. in and around habitations, cultivation, tea gardens, etc. Mainly crepuscular and nocturnal though little incommoded by sunlight, flying freely on disturbance. Pairs spend the daytime secluded in tree-holes or cuddled on a leafy branch. They dash out fussily on suspicion and comically bob and stare at an intruder from a distance. Food: chiefly beetles and other insects; also young mice, lizards, etc. Call: a medley of harsh screeching, chattering and chuckling notes—*chirurrr, chirurrr, chirurrr* ... combined or interlarded with *cheevăk, cheevăk, cheevăk* . . . . Sometimes given as discordant duets.

109 **HIMALAYAN BROWN WOOD OWL**

Strix leptogrammica Temminck

Size Kite − ; dumpier. Length 50 cm (20 in.).

Field Characters A large chocolate-brown owl without ear-tufts. Facial disc whitish, bordered with brown; a prominent white supercilium. Tail brown, barred with fulvous tipped with white. *Below*, buffish, tinged with brown on breast, closely barred with dark brown. A pure white throat-patch. Sexes alike.

Status, Habitat, etc. Resident (subspecies *newarensis*) between 750 and 2450 m—occasionally higher. Affects deep forest. Largely nocturnal.

Plate 14, artist D. M. Henry

MAGPIES, JAY, TREE PIE, PARROTBILL

1 R GREEN MAGPIE, *Cissa chinensis*   page 107
Myna ± . Bright coral-red bill and legs. Up to 1500 m.

2 R REDCROWNED JAY, *Garrulus glandarius*   107
Pigeon − . Broad black moustachial band. Tail black,
not tipped with white. 1500–3000 m.

3 R YELLOWBILLED BLUE MAGPIE, *Cissa flavirostris*   107
Pigeon ± . Crown black. A white nape-patch. Under-
parts primrose-yellow. 1000–3300 m.

4 R BLACKBROWED TREE PIE, *Dendrocitta frontalis*   109
Myna ± . A black hood. Underparts rusty and
whitish. Tail entirely black. Up to 2100.

5 R GREAT PARROTBILL, *Conostoma aemodium*   135
Myna. Stout orange-yellow bill. 2700–3600 m.

HENRY
'53

0   4   8   12 cm

K.P.JADAV

Plate 15, artist K. P. Jadav

FAIRY BLUEBIRD, MINIVET, BULBULS, CHLOROPSIS, BABBLERS

Keeps in pairs to seclusion of dense foliage during daytime, but is capable of flying in bright sunshine without difficulty if disturbed. Food: chiefly birds, often as large as junglefowl, and small mammals. Call: a mellow musical *tok* ... *tū-hoo* (*tok* in undertone, audible only at short range).

110 HIMALAYAN WOOD OWL *Strix aluco* Linnaeus   p. 32

Size Jungle Crow ± ; dumpier. Length 50 cm (20 in.).

Field Characters A medium-sized dark brown owl with a whitish facial disc and no ear-tufts. *Above*, grey-brown, barred with dark brown and mottled with whitish. *Below*, white, closely barred with dark brown on chin and throat; streaked and more openly barred on rest of underparts. Sexes alike.

Distinguished from Brown Wood Owl (109) by smaller size, grey-brown (*v.* chocolate-brown) upperparts, and absence of prominent white supercilium and throat-patch. But has two distinct colour phases (1) more rufous, (2) more greyish rufous.

Status, Habitat, etc. Resident (subspecies *nivicola*) between 1200 and 4250 m: rocky wooded ravines in oak and conifer forest. Largely nocturnal. Solo or pairs. Spends daytime concealed on some densely foliaged branch, standing upright and sleeked, looking like a snag. Hunts at dusk and after dark. Food: rodents, birds, lizards, large insects, etc. Call: a loud *hoo .. hoo .. hoo-ho-ho-hoo*, the final *hoo* drawn out; sometimes merely a hurried deep low *hu-hoo*.

111 SHORTEARED OWL *Asio flammeus* (Pontoppidan)

Size Pigeon + ; length 40 cm (16 in.).

Field Characters A slim medium-sized owl with two short upstanding blackish 'horns' above the staring yellow eyes. Pale buff overall, heavily streaked with dark brown, and with darker greyish head. Facial disc whitish, bordered by a dark brown ruff. Wings and tail barred rufous and black. *Below*, pale buff, streaked with brown on breast. Sexes alike. In flight, the pointed wings, largely rufous above whitish below, with black tips and a dark bar across each surface, are diagnostic pointers.

Status, Habitat, etc. Irregular winter visitor (October to March) and passage migrant. Low country, duars, foothills, and up to 1400 m: undulating grassland and hillslopes dotted with bushes. Solos or loosely scattered parties. More diurnal than most owls, occasionally even hunting in daytime. Normally settles and roosts on the ground. Flies with deliberate full wing beats. Food: field-rats and -mice, lizards, grasshoppers and other large insects. Call: none recorded; very silent in winter quarters.

FROGMOUTHS: Podargidae

112 HODGSON'S FROGMOUTH
Batrachostomus hodgsoni (G. R. Gray) p. 36

Size Myna + ; length 25 cm (10 in.).

Field Characters An obliteratingly coloured nightjar-like bird with exaggeratedly wide gape and broad swollen horny bill, aptly suggestive of a frog's mouth. Male grey-brown, vermiculated and mottled with white, buff, brown, black and chestnut. Scapulars with broad white patches. Tail with pale and dark mottled cross-bands. Female of same camouflaging pattern, but chestnut-rufous overall instead of grey-brown. A more or less distinct white band across throat. A whitish collar round hindneck.

Status, Habitat, etc. Resident; rare, nocturnal and little known. Affects subtropical evergreen forest between 300 and 1800 m. Spends the daytime in thick jungle perched singly, upright and motionless, on a stump or low branch admirably disguised as a lichen-covered snag. Food: moths, beetles and other large insects, hawked in air or taken on ground or from branches. Call: imperfectly known. Described as a soft rapid *kooroo, kooroo, kooroo,* and variously.

NIGHTJARS: Caprimulgidae

113 HIMALAYAN JUNGLE NIGHTJAR
Caprimulgus indicus Latham p. 37

Size Pigeon − ; length 30 cm (12 in.).

Field Characters A crepuscular and nocturnal bird with enormously widened gape, very short legs, and remarkably concealing plumage— mottled and vermiculated grey, brown, fulvous, black and white. Male with white subterminal spots on 4 outer pairs of tail-feathers. First primary wing-quill with a white patch on inner web near tip; the next three quills with white on both webs. Female without the white tips to tail. Spots on primaries rufous (*v.* white), smaller and only faintly indicated. Field recognition from other nightjar species difficult except by its distinctive calls (see below).

Status, Habitat, etc. Resident, subject to seasonal local and altitudinal movements. Foothills and up to 3300 m: forest glades and sparsely scrubbed hillsides and ravines. Spends the daytime, singly or in pairs, resting along a bough or squatting on the ground, effectively camouflaged by its coloration. Emerges at dusk to hawk winged insects, zigzagging, turning and twisting in the air after them with amazing

Plate 16, artist Robert Scholz

SHRIKE, MINIVETS, ORIOLE, STARE, RUBYTHROAT, ROBINS

1 R TIBETAN GREYBACKED SHRIKE, *Lanius tephronotus*   page 98
   Bulbul + . Head and back grey. No white patch on primaries. Foothills to 4500 m.

2 R SHORTBILLED MINIVET, *Pericrocotus brevirostris*   116
   Bulbul − . ♀ Forehead yellow, crown and mantle grey. All red parts replaced by yellow. Duars to 2400 m.

3 R YELLOWTHROATED MINIVET, *Pericrocotus solaris*   117
   Bulbul − . 1500–3000 m.

4 R MAROON ORIOLE, *Oriolus traillii*   100
   Myna + . Duars to 2000 (usually below 1000) m.

5 ? SPOTTEDWINGED STARE, *Saroglossa spiloptera*   104
   Bulbul ± . 700–1200 m.

6 M RUBYTHROAT, *Erithacus calliope*   191
   Sparrow. ♀ Similar but throat white or pinkish without surrounding black line. Duars to 1500 m.

7 R GOLDEN BUSH ROBIN, *Erithacus chrysaeus*   195
   Sparrow. ♀ Olive above, ochre-yellow below. 1400–4600 m.

8 R WHITEBROWED BUSH ROBIN, *Erithacus indicus*   195
   Sparrow. ♀ Olive-brown above; faint white supercilium. Rufous-ochre below. 2000–4200 m.

Inches

0   4   8

R.SCHOLZ

Inches
0   2   4

Plate 17, artist Robert Scholz

CHLOROPSIS, MINLA, BULBULS, NUTHATCH, CREEPERS

1 R   ORANGEBELLIED CHLOROPSIS, *Chloropsis hardwickii*   page 118
Bulbul. Foothills to 2000 m.

2 R   REDTAILED MINLA, *Minla ignotincta*   152
Sparrow. ♀ Paler; back olive-brown. Duars to 3100 m.

3 R   RUFOUSBELLIED BULBUL, *Hypsipetes mcclellandi*   122
Myna ± . Duars to 2700 m.

4 R   WHITETHROATED BULBUL, *Criniger flaveolus*   121
Myna ± . Duars to 1200 m.

5 R   STRIATED GREEN BULBUL, *Pycnonotus striatus*   121
Bulbul ± . Foothills to 2400 m.

6 R   BEAUTIFUL NUTHATCH, *Sitta formosa*   223
Sparrow ± . White edges to tertials. 330–2100 m.

7 R   MANDELLI'S TREE CREEPER, *Certhia familiaris*   226
Sparrow − . Rump ferruginous. Tail not barred.
Throat and breast white; flanks brown. 1700–4000 m.

8 R   WALL CREEPER, *Tichodroma muraria*   226
Sparrow. Throat white in winter. Duars to snow-line.

agility. Flight completely silent as in owls and other soft-plumaged birds. Food: insects—moths, bugs, beetles, etc. Call: a loud, quick-repeated *chuck-chuck-chuck* ... in unbroken runs of 50 or more *chuck*s. A less rapid *chuckoo-chuckoo-chuckoo* ... in runs of 3 to 14, repeated monotonously for several minutes at a stretch with short pauses.

114 LONGTAILED NIGHTJAR *Caprimulgus macrurus* Horsfield

Size Pigeon ± ; slenderer. Length 35 cm (14 in.).

Field Characters A crepuscular and nocturnal cryptically coloured nightjar. Very similar to 113 including white spots on first 4 primaries, but outer two pairs of tail-feathers broadly tipped whitish, and tarsus fully feathered instead of only partially. Distinguishing in the field with certainty difficult except by its distinctive call.

Status, Habitat, etc. Resident and partially migratory. Submontane tracts, foothills, and locally up to 2200 m: shady wooded nullahs in sal and moist-deciduous forest. Habits and food similar to Jungle Nightjar's. Call: a loud, resonant *chaunk, chaunk, chaunk,* etc. recalling heavy blows of a hammer or adze on a wooden plank in the distance. In runs of 50 *chaunk*s or more, repeated with brief pauses from dusk to dawn, especially during moonlit nights in the breeding season.

115 FRANKLIN'S NIGHTJAR
Caprimulgus affinis monticola Franklin

Size Myna + ; length 25 cm (10 in.).

Field Characters Similar to 113 and 114, but smaller. Male. First four primaries black, mottled at tips, with a broad white band across centre. Tail-feathers buff with black cross-bars; the two outer pairs white except at the mottled tips. Tarsus almost naked. Female. Spots on wing rufous-buff instead of white; outer tail-feathers mottled throughout.

Certainly identified only by its distinctive call (see below).

Status, Habitat, etc. Resident; widespread but rather local. Submontane tracts, foothills and locally up to 1800 and even 2400 m: sparsely scrubbed hillsides and nullahs, and thin jungle often in the neighbourhood of cultivation. Habits and food as in other nightjars. Call: a loud, sharp, penetrating single note *sweesh* or *chwees* as of a whiplash cutting the air. Uttered at 4 or 5 seconds' intervals all night, from a perch as well as in flight.

SWIFTS: Apodidae

116 HIMALAYAN SWIFTLET *Collocalia brevirostris* (Horsfield)
Size Sparrow − ; length 14 cm (5½ in.).
Field Characters A small, slender, slightly fork-tailed brown swift, with narrow pointed wings. *Above*, dark brown with noticeably paler rump. *Below*, greyish brown. Sexes alike. In normal hawking flight momentary interludes of bat-like fluttering usually distinguish it from other co-existing swifts.

   Hand Diagnosis. Upper plumage sooty brown; underparts uniform greyish brown. Tips of downy bases of mantle and rump feathers white. Wings mostly over 125 mm; depth of tail-fork 8–10 mm. Tarsus very short, sparsely feathered.
Status, Habitat, etc. Resident; foothills and up to 3600 m. Gregarious; roosts and nests in colonies in limestone caves and grottoes, clinging upright to rough wall surfaces. Loose rabbles spend the entire day on the wing hawking insects, often wandering long distances during the foraging. Food: dipterous and hymenopterous insects—midges, winged ants, etc. Call: a constant conversational twittering at the roost. A wooden rattle-like note used for echo-location while flying in dark caves.

117 'BLACKNEST' SWIFTLET *Collocalia maxima* Hume
Size Sparrow − ; length 14 cm (5½ in.).
Field Characters Very similar to, and practically indistinguishable from Himalayan Swiftlet (116). Slightly heavier build, proportionately broader wings, and less forked tail suggestive.

   Hand Diagnosis. Tips of downy bases of mantle and rump feathers mainly black. Tail almost square: depth of fork 2–3 mm. Tarsus thickly feathered.
Status, Habitat, etc. Resident (?): eastern Bhutan and Arunachal Pradesh (?), between 2100 and 3900 m. Habits and food as in 116. Call: not recorded; probably as 116. Also utters the wooden rattle-like note for echo-location in dark caves.

118 WHITETHROATED SPINETAIL SWIFT
   Chaetura caudacuta (Latham)   p. 37
Size Bulbul + ; stouter. Length 20 cm (8 in.).
Field Characters A large blackish brown swift with long narrow pointed, bow-shaped wings and short tail. Underwing uniform blackish. *Above*, glossy black except middle of back which is pale whitish brown. *Below*,

chin, throat and under tail-coverts white; rest dark brown with a whitish
patch on each flank (subspecies *nudipes*). Sexes alike.

  Hand Diagnosis. Webs of tail-feathers rounded at tip, the rigid
shafts projecting as spines or needles beyond.

Status, Habitat, etc. Resident; rather uncommon and patchily
distributed—normally between 1250 and 4000 m: neighbourhood of
fissured crags and rock scarps. Reputed to be one of our fastest flying
birds. Keeps on the wing all day in loose parties or flocks, swishing
at tremendous speed round contours and hawking over alpine pastures
and river valleys, covering enormous distances in the day's foraging.
Roosts colonially in clefts and fissures (possibly also within hollow
tree-trunks) clinging upright to the rough surfaces. Food: flying
insects, chiefly beetles, bugs and ants. Call: loud, shrill, lively 'screams'
uttered on the wing while disporting themselves prior to retiring at
dusk.

119 **ALPINE SWIFT** *Apus melba* (Linnaeus)   p. 37
Size Bulbul + ; stouter. Length 22 cm (8½ in.).
Field Characters A large streamlined sooty brown swift with very long,
narrow, bow-shaped wings. Underparts white, with a brown pectoral
band across breast. Sexes alike.
Status, Habitat, etc. Resident (possibly largely a seasonal visitor to the
higher altitudes); from plains level up to 2500 m. Subject to considerable
local migration and weather-dependent nomadism, in addition to far-
ranging daily foraging movements. Like 118, exceedingly fast on the
wing. Has the characteristic habit of swifts of 'balling' up in the sky at
sunset in a close-packed rabble, whirling, wheeling and tumbling
playfully to the accompaniment of shrill, joyous screams before retiring
to the communal roost in fissures of precipitous cliffs. Food: as in
118—winged insects hawked in the air. Call: short, shrill, tremulous
twittering screams uttered on the wing, as above.

120 **WHITERUMPED SWIFT** *Apus pacificus* (Latham)
Size Sparrow ± ; length 15 cm (6 in.).
Field Characters *Above*, brownish black, with a broad white rump-patch
and deeply forked tail. *Below*, squamated or mottled black and white
('pepper-and-salt'), with whitish chin and throat (subspecies *leuconyx*).
Sexes alike. On a casual sighting would pass for a House Swift (121),
but paler underparts and forked tail diagnostic.
Status, Habitat, etc. Uncertain. Occurs during breeding season (April-
July) between 600 and 3600 m, but nesting in eastern Himalayas not
yet recorded. Given to capricious wide-ranging nomadic movements.

Keeps in small scattered parties, sometimes large flocks, hawking winged insects all day long: usually at great heights in fine weather, at lower levels and close to ground under cloud overcast. Roosts colonially in rock fissures. Food: flying insects—tiny beetles, bugs, ants, termites, etc. Call: not specifically recorded.

121 HOUSE SWIFT *Apus affinis* (J. E. Gray)   p. 37
Size Sparrow; length 15 cm (6 in.).
Field Characters Swallow-like in flight, but with short square tail and long narrow sickle-shaped wings. Overall smoky black with conspicuous white rump and throat (subspecies *nipalensis*). Sexes alike. Could casually be confused with Whiterumped Swift (120), but blacker underparts and square (*v.* forked) tail diagnostic. The rather similar-looking Martins (175, 176 qq.v.) have white underparts and broader wings.
Status, Habitat, etc. Resident; subject to seasonal local and altitudinal movements. Plains level up to 2000 m: neighbourhood of villages, dzongs and cliffs. Gregarious: habits typical of swifts. Scattered rabbles spend most of the day on the wing, dashing about at great speed on rapidly quivering or stiffly held wings, alternated with swooping glides and agile wheeling and banking movements in pursuit of prey. Roosts at night among clustered colonies of old nests—the 'nest villages'—built within dzongs or on cliffs. Food: tiny winged insects. Call: shrill, joyous, musical twittering 'screams', usually uttered when 'balling' up in the sky at dusk preparatory to roosting.

TROGONS: Trogonidae

122 REDHEADED TROGON *Harpactes erythrocephalus* (Gould) p. 36
Size Myna + ; with longer tail. Combined length 35 cm (14 in.).
Field Characters A brilliantly coloured forest bird. Male. Head, neck and breast deep crimson, with a rudimentary white breast-band. Back rusty brown, vermiculated black and white on wing-coverts and tertiaries. Tail black and white. *Below*, from breast down lighter crimson. Female. Head, neck and breast orange-brown. Rest as in male (subspecies *hodgsoni*).
Status, Habitat, etc. Resident; duars, foothills and up to 1800 m: dense evergreen jungle. Singly or pairs; silent, sluggish, and rather crepuscular. Perches upright on tree stumps or low branches along a shady jungle path or glade and makes fussy, fluttering aerial sorties thence after passing insects, flying on to a fresh perch after each capture. Food:

grasshoppers, cicadas, etc. and larvae. Also berries. Call: an occasional abrupt mewing *cue* (of rich oriole quality) repeated deliberately and unhurriedly 5 or 6 times or more.

123 WARD'S TROGON *Harpactes wardi* (Kinnear)   p. 44

Size Pigeon ± ; with longer tail. Overall length 40 cm (16 in.).

Field Characters A brilliantly coloured rather sluggish forest bird, similar to Redheaded Trogon (122) but larger. Belly crimson-pink in male; primrose-yellow in female. Tail graduated, the feathers squarely truncated at tip. Central rectrices black; lateral ones—pink in male, yellow in female—conspicuous in flight in shady forest.

Status, Habitat, etc.   Resident in the eastern Himalayas, between 1500 and 3000 m, from central Bhutan eastward: lower storey and evergreen undergrowth and bamboo in subtropical forest. Singly or separated pairs. Habits as in 122, q.v. Food: large insects and berries. Call: one described as 'a soft *kew-kew-kew-tiree* uttered at intervals'. Usually silent.

KINGFISHERS: Alcedinidae

124 HIMALAYAN PIED KINGFISHER
  Ceryle lugubris (Temminck)   p. 44

Size House Crow ± ; length 40 cm (16 in.).

Field Characters A large, crested, black-and-white kingfisher with cross-barred upperparts; a broad white nuchal collar. Wings and tail blackish grey, barred and spotted with white. *Below*, white. A broad breast-band of black and rufous-brown spots. Flanks and under tail-coverts barred with blackish. In bright sunlight looks dark bluish grey-and-white rather than black-and-white. Female similar but with pale rust-coloured underwing, clearly diagnostic in flight.

Status, Habitat, etc.   Resident throughout the Himalayan terai, duars and foothills, and locally up to 2000 m. (Paler subspecies *continentalis* from Kashmir to Sikkim and W. Bhutan; darker *guttulata* thence eastward through Arunachal Pradesh.) Affects rocky streams and torrents in the foothills. Pairs perch upright on rocks in a favourite beat of river day after day, bobbing head, erecting crest and jerking tail cocked from time to time. Usually plunges at oblique angle to seize prey near surface, not diving vertically from hovering stance in air like its smaller lowland congener (125). Flight stately, with deliberate wing beats, close over water. Food: fish. Call: an occasional single sharp *click*. Rarely also a loud harsh grating sound, rapidly repeated.

125 **INDIAN PIED KINGFISHER** *Ceryle rudis* (Linnaeus)
Size Myna + ; length 30 cm (12 in.).
Field Characters A speckled and barred black-and-white kingfisher with typical stout dagger-shaped (black) bill. Unmistakable from its spectacular habit of hovering stationary in mid-air and plunging vertically for aquatic prey. Sexes almost alike, but male has a double black breast-band, female only a single.
Status, Habitat, etc. Resident in the duars and lowlands, seldom above 500 m: jheels, irrigation reservoirs, canals, streams, etc. Keeps singly or in pairs, perched on a rock or stake in water, flicking up tail and bobbing, or 'pumping', its head from time to time. Best known for its spectacular mode of hunting. Flies along a few metres above water, bill pointing downward, scanning below for fish near the surface. Checks itself abruptly now and again to investigate closely, 'standing on its tail', poised upright in mid-air on rapidly fluttering wings. Hurls itself headlong upon the quarry, vanishing below the surface soon to reappear with a struggling fish gripped between the mandibles. Food: fish, tadpoles, aquatic insects. Call: a sharp, lively, *chirruk chirruk* uttered chiefly in flight.

126 **GREAT BLUE KINGFISHER** *Alcedo hercules* Laubmann
Size Myna − ; length 20 cm (8 in.).
Field Characters Superficially an enlarged version of the commoner Small Blue Kingfisher; also with short stumpy tail and long, straight, pointed blackish bill. Overall brilliant blue-green above, deep ferruginous (rust-coloured) below excepting white chin and throat. Diagnostic pointers: Crown and nape black closely barred with bluish white. A white stripe down each side of neck. A broad band along middle of back bright pale blue, tail deep blue—very conspicuous in flight. Sexes alike.
Status, Habitat, etc. Resident. Rare: duars and foothills up to 1200 m: shady streams in dense evergreen jungle. Shy and difficult to observe, thus little specifically known. Keeps singly perched on low bushes overhanging a rapid forest stream, plunges on quarry swimming past. Food: fishes and aquatic insects. Call: imperfectly known. Recorded as like the piping *chichee chichee* of the Small Blue while dashing over the water, but louder and less shrill.

127 **BLUE-EARED KINGFISHER** *Alcedo meninting* Horsfield
Size Sparrow; length 16 cm (6 in.).
Field Characters The forest counterpart of the better known Small Blue Kingfisher of opener country. Slightly smaller and much darker: deep

purplish blue above (*v.* bluish green), and with blue ear-coverts instead of rusty (subspecies *coltarti*). Sexes alike.

Status, Habitat, etc. Resident. Uncommon; duars and foothills, normally up to 1000 m, occasionally to 1500 m: dense shady hill-streams in evergreen or heavy bamboo forest. Perches on low herbage overhanging water, bobbing its head and switching up the stumpy tail from time to time. Hunts by plunging vertically upon passing quarry, emerging and flying up with it to the same or a nearby perch where it is battered before being swallowed. Food: tiny fishes and aquatic insects. Call: a shrill *chichee chichee* like Small Blue Kingfisher's but sharper, uttered in flight.

128 THREETOED FOREST KINGFISHER
Ceyx erithacus (Linnaeus)

Size Sparrow ± ; length 13 cm (5 in.).

Field Characters A diminutive resplendent forest kingfisher with bright coral-red bill and feet. *Above*, mantle glistening dark purple-blue or lilac; back and rump brilliant amethyst. A deep blue patch on each side of head. *Below*, bright orange-yellow. A rufous patch on underwing conspicuous in flight. Sexes alike.

Status, Habitat, etc. Resident, subject to considerable local dispersal during rainy season—duars and foothills, normally up to 1000 m: shady jungle streamlets, trickles, puddles, etc. in moist-deciduous and evergreen biotope. Keeps singly, perched quietly on some secluded rock or low branch by water, flashing like a jewel as it dashes away on alarm through the dappled sunshine. Hunts in typical manner of small kingfishers—plunging from perch on quarry sailing past. Food: tiny fishes, crustacea and aquatic insects. Call: a shrill, squeaky *chichee* or *chichichee* while darting off.

129 INDIAN RUDDY KINGFISHER
Halcyon coromanda (Latham)   p. 44

Size Myna + ; length 25 cm (10 in.).

Field Characters A medium-sized cinnamon-coloured kingfisher with a white rump-patch, diagnostic in flight, and bright red bill and feet. *Above*, head, neck and mantle rufous-chestnut or cinnamon, with a glistening red-violet or lilac bloom. Lower back and rump white, tinged with pale blue or violet. *Below*, rufous. Sexes alike.

Status, Habitat, etc. Resident, with little-understood local migrations. Duars, foothills, and up to 1800 m: swamps, shady rivulets and pools in heavy evergreen jungle. Shy and retiring. Keeps singly or in pairs to dense tropical growth. Oftener heard than seen, and habits little known specifically. Food: fish, crustaceans, large insects; maybe lizards and other small creatures like its congeners. Call: a loud, rather musical cackling 'laugh'.

130 WHITEBREASTED KINGFISHER
Halcyon smyrnensis (Linnaeus)

Size Myna + ; length 30 cm (12 in.).

Field Characters A brilliant turquoise-blue kingfisher with deep chocolate-brown head, neck and underparts excepting chin, throat and breast which form a glistening white apron or 'shirt front'. A large white wing-patch, conspicuous in flight. Long, heavy, pointed blood-red bill and legs. Sexes alike.

Status, Habitat, etc. Resident, subject to local movements. Duars and foothills, locally up to 1800 m: wet paddy fields, flooded borrow-pits, ditches, puddles, etc. Less dependent on water than other kingfishers and often found far away from it. Solos or separated pairs. From a favourite lookout stance on telegraph wire or post, pounces on creeping prey and flies off with it to another perch nearby where it is battered and swallowed. Food: fish, frogs, lizards and large insects; occasionally fledgling birds and young mice. Call: a loud cackling 'laugh' uttered from perch and in flight. Song, a long-drawn tremulous musical whistle *kililili* repeated again and again from an exposed tree top.

BEE-EATERS: Meropidae

131 CHESTNUTHEADED BEE-EATER
Merops leschenaulti Vieillot p. 44

Size Bulbul ± ; length 21 cm (8½ in.).

Field Characters A slim grass-green bird with slender, curving, pointed black bill. *Above*, crown, hindneck and upper back bright cinnamon-chestnut. *Below*, chin and throat yellow, bordered by a rufous-and-black gorget; rest of underparts grass-green. Sexes alike.

Status, Habitat, etc. Resident but patchily distributed; subject to local seasonal movements, particularly during the monsoon. Duars, foothills, and locally up to 1500 m: mixed moist-deciduous forest, especially the neighbourhood of streams. Small parties or flocks usually seen on top bare branches of forest trees or telegraph wires, launching out in the air, one or several at a time, to hawk winged insects, and circling back to the base after each capture. Flight swift—a few rapid wing beats followed by a graceful glide. Has communal roosts in leafy trees where large numbers collect for the night. Food: winged insects—bees, wasps, dragonflies, termites, etc. Call: a musical interrogative, repetitive *tetew?*, at times sounding rather like Redvented Bulbul's *pettigrew* call.

132 **BLUEBEARDED BEE-EATER**
Nyctyornis athertoni (Jardine & Selby)   p. 44
Size Pigeon − ; slimmer. Length 35 cm (14 in.).
Field Characters A large green bee-eater with long, slender, slightly curved black bill and square-cut tail, without projecting pin-feathers. *Above*, grass-green with pale bluish forehead. *Below*, upper breast verditer-blue. A light blue 'beard' of elongated feathers on lower throat, conspicuous in profile (when standing away from breast while bird calling). Rest of lower parts rusty buff, broadly streaked with green. Sexes alike.
Status, Habitat, etc. Resident; duars, foothills, and up to 1700 m: overgrown ravines and broken country in secondary evergreen or moist-deciduous forest. Arboreal. Keeps in pairs to leafy tree tops. Launches aerial sorties after flying insects, returning to its base after each capture to batter the victim against the perch before being swallowed. Flight steeply undulating—a few rapid wing flaps followed by a downward glide with wings closed. Food: insects, chiefly bees, wasps, dragonflies and beetles, etc. Call: loud, hoarse guttural croaks and chortles.

ROLLERS or BLUE JAYS: Coraciidae

133 **ROLLER** *Coracias benghalensis* (Linnaeus)   p. 44
Size Pigeon; length 30 cm (12 in.).
Field Characters A striking bright dark-and-light blue bird with biggish clumsy head and crow-like bill. In flight the dark and pale portions of the wings flash as brilliant bands. *Below*, largely rufous-brown, streaked with blue on throat and washed with purplish on breast; vent and under tail-coverts pale blue (subspecies *affinis*). A deep purple-blue patch under wing (coverts), conspicuous when overhead. Sexes alike.
Status, Habitat, etc. Resident, subject to seasonal local movements. Plains, duars and foothills up to 600 m: cultivation, and clearings in light forest. Keeps singly or in pairs perched on telegraph wires, poles, etc., on vigil for creeping prey on the ground. Pounces on quarry and flies back with it lazily to the same or another perch where it is battered and swallowed. Has a spectacular aerial display in breeding season,

rocketing vertically, nose-diving, rolling from side to side on the wing and performing fantastic aerobatics to the accompaniment of loud discordant screams, its brilliant plumage flashing in the sun. Sometimes both sexes participate. Food: large insects, lizards, frogs, etc. Call: a loud, raucous *kāk-kāk-kāk-kāk* etc. uttered from perch or during display.

134 **BROADBILLED ROLLER** *Eurystomus orientalis* (Linnaeus) p. 44
Size Pigeon; length 30 cm (12 in.).
Field Characters A dark purplish brown and blue-black roller, with blackish head and very broad orange-red bill and orange-red legs. A large roundish pale blue patch on wings conspicuous in flight. Sexes alike.
Status, Habitat, etc. Resident (subspecies *cyanicollis*). The entire sub-Himalayas from Garhwal eastward—terai, duars, foothills and up to 1000 m: heavy secondary tropical evergreen jungle and cultivation clearings with tall relict trees. Usually singly or pairs, perched on dead trees as lookout posts. Captures winged insects by aerial sorties, performing agile evolutions in their pursuit and returning to base with the quarry. Also takes creeping prey from the ground. Is less lethargic than common Roller, and markedly crepuscular. Food: chiefly large insects; also lizards and other small animals. Call: raucous *chack-chack* at intervals, occasionally given as a quick-repeated chattering croak.

HOOPOES: Upupidae

135 **HOOPOE** *Upupa epops* Linnaeus
Size Myna; length 30 cm (12 in.) including long slender bill.
Field Characters A fawn-coloured bird with very conspicuous black and white zebra markings above. A prominent full fanlike crest tipped with black and white, which falls into a point behind the head when folded. Long, slender, gently decurved brown bill (subspecies *saturata*). Sexes alike.
Status, Habitat, etc. Resident and/or summer (breeding) visitor in the Tibetan Plateau facies of the High Himalayas above 1700 m, and locally up to 4000 m. Descends to lower levels and the duars and adjoining plains in winter. Affects lightly wooded country and the neighbourhood of upland villages and dzongs; singly or in separated pairs. Feeds on the ground, running about like quail, probing into loose soil for hidden insects. When thus digging, folded crest sticks out in a point behind head looking like a miniature pickaxe. Often indulges in curious erratic

butterfly-like flights. Food: insects, grubs and pupae. Call: a deep, mellow *hoo-po* or *hoo-po-po* casually confusable with call of Himalayan Cuckoo, *Cuculus saturatus* (88) q.v.

HORNBILLS: Bucerotidae

136 RUFOUSNECKED HORNBILL
Aceros nipalensis (Hodgson)   p. 45

Size Vulture; length 120 cm (48 in.).

Field Characters Easily identified by large size and distinctive colour pattern. Male. *Above*, head, neck and breast rufous; rest of upperparts glistening black; terminal half of tail white. *Below*, rufous, maroon and black. Female. Entirely black except tips of outer wing-quills and terminal half of tail which are white as in male.

Status, Habitat, etc. Resident. Himalayan foothills from Nepal eastward, up to 1800 m: tall subtropical evergreen forest. Pairs, small parties, or larger feeding flocks keeping to lofty trees, or seen sailing majestically across wooded valleys. Flight: a few rapid, noisy flaps followed by a long glide on flat outstretched wings ending in a graceful upward sweep to settle on a branch. Food: large drupes and berries, occasionally picked off the ground in ungainly shuffling hops with tail partly cocked. Call: a variety of loud far-carrying roars, croaks and cackles; frequently uttered by a pair as an uneven duet.

137 WREATHED HORNBILL *Rhyticeros undulatus* (Shaw)   p. 45

Size Vulture ± ; overall length ♂ 110 cm (44 in.), ♀ 100 cm (40 in.).

Field Characters A large black hornbill with entirely white tail. Male. Forehead, crown and nape deep chestnut passing into black on hindneck and into whitish on foreneck and sides of head. Tail entirely white. Rest of plumage, above and below, glossy black. Bill pale yellow with reddish corrugations at base of both mandibles; a bright yellow throat-patch. Female entirely black with entirely white tail. Colour pattern of both sexes diagnostic even in high overhead flight.

Status, Habitat, etc. Resident. Duars and foothills up to about 2400 m, but chiefly lowland: dense tropical evergreen primeval forest. Keeps in pairs or small parties; larger congregations on fruiting trees and at nightly roosts in favourite groves of giant bamboo or tall thinly foliaged trees. Habits as of other hornbills; arboreal and mainly frugivorous. Food: large drupes and berries; lizards and other small animals. Call: short raucous grunts uttered from a perch, repeated several times.

138 PIED HORNBILL *Anthracoceros malabaricus* (Gmelin)

Size Vulture − ; length ♂ 90 cm (36 in.).

Field Characters A large black-and-white hornbill with black neck, white underparts, and ponderous pale yellow and black bill surmounted by a ridge-like casque ending in a single point in front. Distinguished from Great Pied Hornbill (139) by smaller size, differently shaped casque, q.v., black neck (*v.* white), and broadly white-ended black outer rectrices (*v.* all-white tail with a broad black cross-band). Sexes alike; female slightly smaller.

Status, Habitat, etc. Resident, subject to seasonal local movements. The Himalayas from about Dehra Dun eastwards—terai, bhabar, duars and foothills, up to 600 m: open subtropical moist-deciduous and evergreen forest. Sociable and arboreal. Noisy parties of 4 or 5 and small flocks; large gatherings on fruiting *Ficus* trees in association with other frugivorous birds. Flight: a few noisy flaps followed by a glide with upturned wing-tips. Food: mainly wild figs, drupes and berries; also lizards, snakes, and sundry small animals. Call: loud shrill squeals and raucous cackles.

139 GREAT PIED HORNBILL *Buceros bicornis* Linnaeus   p. 45

Size Vulture; length 130 cm (52 in.).

Field Characters Large size, black face, white neck, and enormous bill with concave-topped casque ending in two points at front, readily distinguish it from the smaller Pied Hornbill (138). Black breast, and entirely white tail with a single black band, additional pointers. In flight, white neck, broad white wing-bar and black-banded all-white tail diagnostic. Female similar but smaller.

Status, Habitat, etc. Resident, subject to seasonal local movements: the Himalayas from Kumaon eastwards—terai, bhabar, duars and foothills, up to 2000 m: primeval subtropical evergreen and moist-deciduous forest. Sociable and arboreal, with habits as in 138, with which its range largely overlaps in eastern Himalayas. Gatherings of 150 or more at profusely fruiting *Ficus* trees in company with other frugivorous birds. Flight as in 138; the loud scraping sound produced by the wings audible at a great distance. Roosts colonially among thinly foliaged top branches of lofty trees. Food: wild figs, nutmegs, drupes and berries; also miscellaneous animal items, such as lizards, snakes, rodents, young birds. Call: deep hoarse resonant grunts, and loud reverberating 'barks'.

BARBETS: Capitonidae

140 GREAT BARBET *Megalaima virens* (Boddaert)   p. 48

Size Myna + ; length 35 cm (14 in.).

Field Characters A gaudy arboreal bird with heavy ungainly yellow bill surrounded at base by prominent bristles. *Above*, maroon-brown with violet blue-black head. *Below*, olive-brown, blue, and yellow, with a bright scarlet patch under tail (coverts). Sexes alike. Dumpy build and dipping woodpecker-like flight with expanded triangular tail identify it as a barbet a long way off.

Status, Habitat, etc. Resident, subject to winter-summer altitudinal movements. Himalayas from W. Nepal eastward through Sikkim, Bhutan and Arunachal Pradesh, between 1000 and 3000 m (subspecies *magnifica* and *mayri*): subtropical and temperate evergreen and moist-deciduous forest. Keeps singly or in small parties to tall trees on hillsides—larger gatherings where food abundant. Despite its gaudy coloration merges astonishingly into the foliage when sitting motionless among upper branches, and difficult to see. Very vociferous during summer when its all-pervading calls resound through the wooded valleys. Flight swift and undulating, a few rapid flaps followed by a dipping glide with closed wings. Food: fruit, flower petals and insects—sometimes captured in clumsy aerial sorties. Call: a mournful wailing *piāo* or *pihow* repeated monotonously in long runs throughout the day.

141 LINEATED BARBET *Megalaima lineata* (Vieillot)   p. 44

Size Myna + ; length 30 cm (12 in.).

Field Characters A dumpy, stout-billed and 'whiskered' grass-green arboreal bird. Head, neck and upper breast brown, coarsely streaked with whitish; a naked yellow patch round eye (not extending to gape). Sexes alike.

Status, Habitat, etc. Resident. Duars and foothills, up to 800 m: moist-deciduous biotope (subspecies *hodgsoni*). Frequents light secondary forest, well-wooded gardens, roadside avenues, etc. Singly or in small feeding parties on large fruiting *Ficus* and suchlike trees in association with other frugivorous birds. Extremely noisy in hot weather, the countryside then ringing with its incessant calling. Flight typical of barbets—heavy, noisy and dipping. Food: drupes, berries, flower petals and nectar; also insects and occasionally lizards etc. Call: monotonous and almost non-stop, beginning with a harsh *krr, krr* and settling down to an unvarying *kotur, kotur, kotur* ... (or *pocock, pocock, pocock* ...) all individuals within earshot joining in an uneven chorus.

142 **GOLDENTHROATED BARBET**
Megalaima franklinii (Blyth) p. 48
Size Myna; length 23 cm (9 in.).

Field Characters A dumpy grass-green barbet with crimson and bright golden crown, grey ear-coverts, and a broad black stripe above eye from bill to nape. Chin and throat golden yellow; an orange spot on each side of base of bill near gape. Sexes alike.

Status, Habitat, etc. Resident between 600 and 2400 m. In summer found at higher elevations than Bluethroated Barbet (next): forested hillsides in subtropical moist-deciduous biotope (subspecies *franklinii*). Frugivorous, with typical barbet habits and behaviour (cf. 140, 141). Call: calling begins with a rolling *krrr-krrr* and settles down to a monotonous *pūkwowk, pūkwowk, pūkwowk* reiterated in intermittent runs of many minutes each, *ad nauseam*. The initial *pūk* is audible only at short range, thus in the distance the call sounds more like a single-noted *wowk, wowk, wowk* (*ow* as in owl).

143 **BLUETHROATED BARBET** *Megalaima asiatica* (Latham) p. 44
Size Myna; length 23 cm (9 in.).

Field Characters A gaudily-coloured barbet, chiefly grass-green, with the typical heavy conical bill. Forehead and forecrown crimson, then yellowish, followed by a transverse black band above the eyes. Hindcrown crimson with a black streak on either side running back to nape. A short supercilium, feathers round eye, ear-coverts, chin and throat verditer blue. A crimson spot on each side of base of lower mandible; a crimson patch at base of throat on either side. Base of bill surrounded by conspicuous black bristles or whiskers. Sexes alike. Some examples have the green upperparts suffused with crimson and under-parts also flecked and streaked with crimson in varying degree. (Untenably described by Baker as the subspecies *rubescens*.)

Status, Habitat, etc. Resident, from plains level to 2000 m, occupying a lower altitudinal zone in summer than Goldenthroated Barbet (142). Affects well-wooded country: light deciduous to evergreen forest, and groves of *Ficus* and suchlike fruit-bearing trees around villages and in urban gardens. Habits, food, etc. as of other barbets (cf. 140, 141). Call: almost indistinguishable from that of Lineated Barbet (141), but softer, higher in key and somewhat quicker in tempo. Sounds more like *pūkūrūk, pūkūrūk* (3 syllables) than *pocock* or *kotur* (2 syllables) of Lineated; commonly given in irregular duets or choruses.

144 **BLUE-EARED BARBET** *Megalaima australis* (Horsfield)
Size Sparrow + ; length 17 cm (6½ in.).

Plate 18, artist J. P. Irani

CHOUGHS, MAGPIE, LAUGHING THRUSHES, SCIMITAR BABBLERS

1 R **ALPINE CHOUGH,** *Pyrrhocorax graculus*   page 110
House Crow – . Like 2, but bill shorter, yellow. Legs bright
red. 2400–5000 m.

2 R **REDBILLED CHOUGH,** *Pyrrhocorax pyrrhocorax*   110
House Crow ± . Longish slightly curved bright red bill, and
bright red legs. 1600–4500 m.

3 R **BLACKRUMPED MAGPIE,** *Pica pica bottanensis*   108
Myna + . 3400–4600 m.

4 R **WHITESPOTTED LAUGHING THRUSH,** *Garrulax ocellatus*   142
Pigeon. Crown and throat black. 2100–3400 m.

5 R **CRIMSONWINGED LAUGHING THRUSH,**
Garrulax phoeniceus   147
Myna ± . Duars to 1800 m.

6 R **CORALBILLED SCIMITAR BABBLER,** *Pomatorhinus*
ferruginosus   126
Bulbul + . Bright coral-red bill. Underparts ferruginous. Duars
to 3800 m.

7 R **RUFOUSCHINNED LAUGHING THRUSH,**
Garrulax rufogularis   141
Myna. Tail chestnut with black subterminal band. 600–1900 m.

8 R **RUFOUSNECKED SCIMITAR BABBLER,**
Pomatorhinus ruficollis   125
Bulbul ± . Rufous patch on each side of neck. 700–3000 m.

9 R **STRIATED LAUGHING THRUSH,** *Garrulax striatus*   140
Myna + . Mop-like crest. Low foothills up to 2700 m.

10 R **BLUEWINGED LAUGHING THRUSH,** *Garrulax squamatus*   143
Myna + . Rufous shoulder-patch. Rufous terminal band to
blackish tail. 1000–2400 m.

11 R **SLENDERBILLED SCIMITAR BABBLER,**
Xiphirhynchus superciliaris   127
Bulbul. Very long slender curved bill. 600–3400 m.

12 R **REDHEADED LAUGHING THRUSH,** *Garrulax*
erythrocephalus   146
Myna ± . Wings olive-yellow. Chestnut shoulder-patch. 1200–
3000 m.

13 R **GREYSIDED LAUGHING THRUSH,** *Garrulax caerulatus*   142
Myna ± . White underparts; grey flanks. 1500–2700 m.

D.M.H.
1955

Plate 19, artist D. M. Henry

BABBLERS, SIVA, SIBIA, BARWING, CUTIA, LEIOTHRIX

1 R   REDCAPPED BABBLER, *Timalia pileata* page 134
Sparrow +. Narrow white forehead continued as short supercilium. Throat white; sides of neck grey. Duars and foothills.

2 R   WHITEHEADED SHRIKE-BABBLER,
Gampsorhynchus rufulus 150
Bulbul +. White shoulder-patch on wing. Duars to 1200 m.

3 R   SPOTTED BABBLER, *Pellorneum ruficeps* 123
Bulbul –. Duars to 1800 m.

4 R   BARTHROATED SIVA, *Minla strigula* 152
Sparrow. 1800–3600 m.

5 R   LONGBILLED WREN-BABBLER, *Rimator malacoptilus* 127
Sparrow ±. Long, slender, slightly curved black bill. Stub tail. 900–2700 m.

6 R   REDWINGED SHRIKE-BABBLER, *Pteruthius flaviscapis* 149
Myna –. ♀ Head and back grey. Wings chestnut and black, edged yellowish green. Underparts buff. 1500–2500 m.

7 R   RUFOUSBELLIED SHRIKE-BABBLER, *Pteruthius rufiventer* 149
Bulbul. ♀ Head grey. Back and mantle greenish yellow. Secondaries tipped chestnut. 1500–2500 m.

8 R   CHESTNUTBACKED SIBIA, *Heterophasia annectens* 157
Bulbul. Crown black. Rump chestnut. Underparts white. Foothills to 2300 m.

9 R   HOARY BARWING, *Actinodura nipalensis* 151
Bulbul. Dark brown crown, grey breast, blackish tail. 2100–3300 m.

10 R   CUTIA, *Cutia nipalensis* 148
Bulbul. ♀ Eye-band chocolate brown. Back rufous with oval black spots. 1350–2500 m.

11 R   REDBILLED LEIOTHRIX, *Leiothrix lutea* 148
Sparrow –. ♀ Crimson of wings replaced by yellow. 1500–2400 m.

Field Characters A small bright grass-green barbet distinguished by its gaudy multicoloured head-pattern. 'Whiskers' (rictal bristles) projecting beyond bill tip. Forehead and forecrown black, the feathers fringed with pale blue; hindcrown cobalt blue. Ear-coverts pale verditer blue with a crimson patch above and another below. Chin and throat pale verditer blue. A patch below eye mixed yellow and bright scarlet, separated from chin by a black moustachial stripe. Sexes alike.

Status, Habitat, etc. Resident. Duars and foothills up to 1200 m (subspecies *cyanotis*): thick evergreen forest. Occurrence in Arunachal Pradesh needs confirming. Closely resembles Crimsonbreasted Barbet or Coppersmith (next) in habits and behaviour; just as noisy, but more restricted to heavy jungle. Call: a peculiar, somewhat harsh metallic double note *ko-tūrr* or *too-rook*, endlessly repeated from a tree top. Unlike that of any other barbet, but closest in pattern to that of Lineated.

145 CRIMSONBREASTED BARBET or COPPERSMITH
Megalaima haemacephala (P. L. S. Müller)

Size Sparrow + ; length 17 cm (6½ in.).

Field Characters A small dumpy grass-green barbet with yellow throat, crimson breast and forehead, and green-streaked yellowish underparts. Short truncated tail distinctly triangular in flight silhouette. Sexes alike.

Status, Habitat, etc. Resident. Duars and foothills up to 2000 m: dry- and moist-deciduous biotope—light forest, groves of *Ficus* and other fruiting trees about villages and in urban gardens. Mainly frugivorous: occasionally takes insects. Singly or pairs; occasionally parties of a dozen or more on fruit trees in company with mynas, bulbuls and such other birds. Rather silent in winter; obstreperous as the season warms up. Call: a loud, rather metallic *tūk ... tūk ... tūk* and so on, reminiscent of a distant coppersmith hammering on his metal, or the time-pips of All India Radio. With each *tūk* the head is bobbed and turned from side to side, producing a curious ventriloquistic effect.

HONEYGUIDES: Indicatoridae

146 ORANGERUMPED HONEYGUIDE
Indicator xanthonotus Blyth   p. 48

Size Sparrow; length 15 cm (6 in.).

Field Characters A sparrow-like dark olive-brown bird with glistening orange-yellow forehead and cheeks. Middle of back pure yellow, lower back and rump orange-yellow, conspicuous in the characteristic sitting position with partly drooping wings; also prominent in flight. Female nearly the same: somewhat duller coloured.

Status, Habitat, etc. Presumably resident, between 1500 and 3500 m (subspecies *xanthonotus* and *fulvus*). Very rare and sporadic, and little known. Affects forest in the neighbourhood of bee-nesting cliffs. Food: bees, and probably other hymenoptera; taken in air or while clinging to honeycombs like a woodpecker. Also flaky wax of disused combs. Call: unknown except an occasional single *weet* uttered during the aerial sallies after insects.

WOODPECKERS: Picidae

147 WRYNECK *Jynx torquilla* Linnaeus
Size Bulbul − ; length 19 cm (7½ in.).
Field Characters A slim, silvery grey-brown bird, deceptively sparrow-like particularly in flight. *Above*, streaked, speckled and vermiculated with black and fulvous. *Below*, whitish with blackish arrow-head markings producing a finely cross-barred pattern. Tail with 3 or 4 conspicuous dark bands. Sexes alike.
Status, Habitat, etc. Winter visitor, *c.* September to March-April, chiefly terai and duars: open scrub country and cultivation environs. Singly or pairs; easily overlooked because of its unobtrusive behaviour and remarkably concealing coloration. Hops on ground with tail slightly cocked to pick crawling insects. Clings to upright stems like a true woodpecker, but also perches across a branch like a passerine bird. Dipping flight and general behaviour very finch-like. When handled, stretches neck and bill upright and screws neck comically from side to side like some clockwork toy, whence its name. This is presumably a threatening gesture. Food: mainly ants. Call: a shrill, quick-repeated, rather nasal *chewn, chewn, chewn* which usually betrays its presence in a locality.

148 SPECKLED PICULET *Picumnus innominatus* Burton
Size Sparrow − ; length 10 cm (4 in.).
Field Characters A diminutive woodpecker with short, soft, rounded black-and-white tail. *Above*, bright yellowish olive. Forecrown black and orange; a broad blackish olive band behind eyes and down sides of neck, bordered above and below by conspicuous whitish bands; a dark moustachial stripe under the lower band. *Below*, yellowish white with bold black spots turning into bars on posterior flanks. Female has the whole crown yellowish olive, uniform with back.

Status, Habitat, etc. Resident. Duars, foothills and up to 2000 m: secondary jungle and tangled brushwood with bamboos—moist-deciduous and semi-evergreen biotope. Singly or pairs, often among the mixed hunting parties of leaf warblers and small babblers where apt to be overlooked. Creeps jerkily along or in spirals round thin end-twigs of low trees and bushes like nuthatch, tapping on the bark now and again in true woodpecker fashion. Also perches across like a passerine bird. Flight strong and direct. Food: mainly ants and their eggs. Call: a sharp *spit, spit* frequently repeated. Also a persistent mechanical drumming *brr-r-r, brr-r-r* on a bamboo stem or dead snag.

149 RUFOUS PICULET *Sasia ochracea* Hodgson p. 48
Size Sparrow − ; length 9 cm (3½ in.).
Field Characters A diminutive, dumpy, stub-tailed woodpecker very different from the conventional shape. Male. Deep rufous-olive with rufescent golden forehead, a short broad white stripe above and behind eye, and stumpy black tail. Female similar but lacking the golden forehead.
Status, Habitat, etc. Resident. Duars, foothills, and locally up to 2100 m. Not uncommon, but unobtrusive and easily overlooked. Singly or pairs, often in association with mixed hunting parties of warblers, small babblers, etc. Habits similar to Speckled Piculet (148), q.v. Sometimes hops amongst mulch on the forest floor with its stub tail cocked like a wren's. 'Drumming' has not been recorded. Food: mainly ants and their eggs and pupae. Call: a miniature replica of the querulous scream-ing as of the larger woodpeckers, uttered in flight as well as while creeping up stems and twigs.

150 RUFOUS WOODPECKER *Micropternus brachyurus* (Vieillot)
Size Myna ± : length 25 cm (10 in.).
Field Characters A chestnut-rufous woodpecker, narrowly cross-barred with black on upperparts, wings and tail. Feathers of throat pale-edged producing a streaked or scaly pattern. Male distinguished from female by a crescent-shaped patch of crimson feathers under eye.
Status, Habitat, etc. Resident. Duars, and foothills up to 1500 m (commoner below 700 m)—in moist-deciduous biotope (subspecies *phaioceps*): secondary forest mixed with bamboo—especially where abounding in carton nests of tree ants. Singly or pairs; frequently seen clinging on and digging into the papiermâché-like nests to feed on the insects. Normally breeds in hollows excavated in such nests even when occupied, apparently enjoying complete immunity from the attacks of the ferocious insects. Rather silent vocally but much given

to 'drumming'. Food: mainly ants and their pupae. Call: a high-pitched nasal *keenk* .. *keenk* repeated 3 or 4 times, very similar to one of the commoner 'shrieks' of the Indian Myna.

151 **BLACKNAPED GREEN WOODPECKER** *Picus canus* Gmelin
p. 45

Size Pigeon ± ; length 30 cm (12 in.).

Field Characters A largish woodpecker, dark green above, tinged on rump with bright yellow. Forehead and forecrown crimson; hindcrown, occipital crest and nape black. Sides of head and supercilium grey, the latter bordered above by a black line. Wing-quills blackish, barred with white; tail black with imperfect pale bars. *Below*, chin and throat grey; a black malar stripe from lower mandible down each side. Rest of underparts dull yellowish green, greyer on belly. **Female** similar but with no crimson on forehead and forecrown which are black like hindcrown; nape streaked with grey.

Status, Habitat, etc. Resident. Duars, foothills and up to 2100 m, perhaps higher (subspecies *gyldenstolpei*; possibly also *kogo* in NE. Arunachal Pradesh): open mixed forest with bamboo, in semi-evergreen biotope. A typical woodpecker; normally seen clinging to stems and branches of trees. Works upward in spurts, directly or in spirals, tapping on the bark from time to time to dislodge insects lurking in crevices or to locate beetles' borings. Sometimes descends to ground and hops about picking ants etc. Flight swift and bounding: 4 or 5 rapid wing flaps followed by a long dipping glide. Food: insects, especially ants and larvae of wood-boring beetles; sometimes a few berries. Call: a high-pitched rather musical *peek* ... *peek* ... repeated unhurriedly in runs of 4 or 5; also a loud mechanical drumming on wood in series of bursts 2 or 3 seconds long each.

152 **LARGE YELLOWNAPED WOODPECKER**
Picus flavinucha Gould
p. 48

Size Pigeon ± ; length 35 cm (14 in.).

Field Characters A largish yellow-green woodpecker with rufous-banded dark brown wings; a conspicuous golden yellow nuchal crest, erected fanwise under excitement. Underparts olive-grey. Tail black, unbarred. **Male** has chin and throat bright lemon-yellow; foreneck rich brown streaked with white. **Female** similar but with chin and throat rufous-brown instead of yellow.

Status, Habitat, etc. Resident. Duars, foothills and locally up to 2400 m; preferential zone between 700 and 1500 m: open mixed evergreen and deciduous forest; partial to edge of cultivation clearings and broken foothills country. Habits and behaviour typical of the family. Pairs, or parties of 4 or 5, often in association with racket-tailed and other drongos. Food: ants, termites, grubs of wood-boring beetles, etc. Call: a loud, plaintive *pee-ū .. pee-ū*; a single metallic contact note *chĕnk*; a rich musical chattering 'laugh' when flying off on disturbance.

153 SMALL YELLOWNAPED WOODPECKER

Picus chlorolophus Vieillot

Size Myna + ; length 25 cm (10 in.).

Field Characters A medium-sized yellowish green woodpecker with golden yellow nuchal crest. Male. *Above*, bright yellowish green; wing-quills largely green and maroon-red. Forehead, supercilia (continued behind to meet at nape) and a moustachial streak crimson; crest golden yellow. Tail blackish and bronze-green. *Below*, chin and throat brown, barred with whitish; breast olive-brown; rest barred brown and white. Female similar but crimson on head restricted to a short broad line behind eye to nape.

Status, Habitat, etc. Resident. Duars and foothills, locally up to 2000 m: mixed deciduous and evergreen secondary jungle. Typical woodpecker. Pairs, often amongst the mixed itinerant hunting parties. Food: insects, chiefly ants, termites and beetle larvae. Call: a long-drawn plaintive nasal contact note *cheenk*, repeated monotonously every 15 to 30 seconds for long periods from the same spot. A distinctive trill of 5 or more ascending notes mistakable for some kind of cuckoo's, and a loud *quaaa* at intervals.

154 GOLDENBACKED THREETOED WOODPECKER

Dinopium shorii (Vigors)   p. 48

Size Pigeon ± ; length 30 cm (12 in.).

Field Characters A largish goldenbacked woodpecker with crimson rump and black tail. Male. *Above*, crown and occipital crest crimson. Hindneck black, the black continued forward as a black stripe to behind eye; a prominent white supercilium to nape; a broad white band down each side of neck. Black moustachial streaks, continued as a double line down middle of throat with the intervening space pale brown. *Below*, buffy white boldly streaked and scalloped with black. Female similar but with forehead and forecrown blackish instead of crimson; crown and crest black with long white streaks.

Status, Habitat, etc. Resident. Duars and foothills up to 700 m, narrowly restricted to tall deciduous and semi-evergreen forest, thus rather patchily distributed. Habits, behaviour and food typical of the family.

Call: a loud high-pitched cackling 'laugh', rather like the Whitebreasted Kingfisher's but more tinny.

155 PALEHEADED WOODPECKER *Gecinulus grantia* (Horsfield)

p. 45

Size Myna + ; length 25 cm (10 in.).

Field Characters Male. *Above*, largely dull crimson. Forehead and sides of head pale golden olive-brown; a crimson-pink patch on crown. *Below*, chin and throat dull olive-yellow merging into dark brownish olive of rest of underparts. Female merely lacks the crimson-pink patch on crown.

Status, Habitat, etc. Resident; not uncommon but rather local. Duars and foothills up to 1000 m, occasionally higher: bamboo and mixed secondary forest in moist-deciduous biotope. Very partial to bamboo jungle. Keeps singly or in pairs, hunting on moderate-sized tree-trunks and large bamboo culms. Also on fallen decaying boles, and rarely on the ground. Food: chiefly ants and grubs of beetles. Call: a loud, harsh, castanet-like rattling *kĕrĕkē-kĕrĕkē-kĕrĕkē* ..., reminiscent of a tree pie or jay. A nasal *chaik-chaik-chaik* ..., five or six times, starting loud and slow, growing faster and fainter and fading out. The first apparently when agitated, the second the normal contact call.

156 GREAT SLATY WOODPECKER

Mulleripicus pulverulentus (Temminck)

p. 48

Size House Crow + ; length 50 cm (20 in.).

Field Characters A large woodpecker, overall slaty grey with buffy yellow chin, throat and foreneck. Male with short broad crimson moustachial stripe; female without.

Status, Habitat, etc. Resident. Duars, and foothills up to 1000 m, occasionally to 2000 m (subspecies *mohun* and *harterti*): climax sal and tropical semi-evergreen forest, and overgrown clearings with a sprinkling of tall relict trees. Normally scattered parties of 3 to 6 which maintain contact by short querulous notes and intermittent powerful tapping while scuttling jerkily up and around the boles. Flight crow-like and seemingly leisurely, without the characteristic woodpecker bounds. Food: insects, chiefly wood-boring beetle larvae. Call: a loud raucous cackle in flight. A reverberating, far-carrying mechanical drumming with bill on tree-trunks.

157 RUFOUSBELLIED WOODPECKER

Hypopicus hyperythrus (Vigors)

Size Bulbul ± ; length 20 cm (8 in.).

Field Characters A practically uncrested black-and-white and chestnut woodpecker. **Male**. *Above*, broadly barred black and white. Wings and tail black, largely spotted and barred white; crown and nape crimson. *Below*, chin grey; throat, sides of neck and underparts bright chestnut; under tail-coverts pale crimson. **Female** similar but crown and nape black, spotted with white.

Status, Habitat, etc. Resident: in pine, moist temperate and subtropical forest—rhododendron, oak, horse-chestnut, etc. up to 4000 m (subspecies *hyperythrus*). Singly or separated pairs, commonly in association with the itinerant hunting parties of small insectivorous birds, in low bushes as well as tall trees. **Food**: ants, grubs of beetles and other insects. **Call**: described as a long but not loud rattling cry ending like the rapid running out of a large fishing reel! Mechanical drumming on dead wood by both sexes.

158 DARJEELING PIED WOODPECKER
Picoides darjellensis (Blyth)   p. 48

Size Myna $\pm$; length 25 cm (10 in.).

Field Characters A medium-sized high-elevation pied woodpecker. **Male**. *Above*, back and middle tail-feathers entirely black; outer barred with white. Hindcrown and nape crimson. *Below*, yellowish fulvous streaked with black; vent and under tail-coverts pale crimson. **Female** similar but with hindcrown and nape black.

Status, Habitat, etc. Resident between 1700 and 3500 m even in winter; in summer up to 4000 m: pine, oak, rhododendron and subtropical wet forest. Singly or pairs. Hunts high up on moss-covered branches of tall trees amongst the canopy foliage as well as on storm-blown trunks littering the forest floor. **Food**: larvae of beetles, and other insects. **Call**: poorly known; described as 'an occasional low *puk* ... *puk*' in addition to the customary drumming of the family.

159 CRIMSONBREASTED PIED WOODPECKER
Picoides cathpharius (Blyth)   p. 45

Size Bulbul − ; length 18 cm (7 in.).

Field Characters A smallish high-elevation woodpecker, a passable miniature of the Darjeeling Pied, q.v. **Male**. *Above*, back and tail black, the latter with outer feathers barred whitish. Crimson of hindcrown and occipital crest extends to entire sides of neck. *Below*, streaked with black as in Darjeeling Pied, but underparts darker and more fulvous, with a diffuse crimson patch on breast. Under tail-coverts fulvous, streaked with black, edged and tipped crimson. **Female** similar but with entire crown black; sides of neck whitish faintly tinged with crimson.

Status, Habitat, etc. Resident, with a more or less overlapping altitudinal range with Darjeeling Pied (158) (subspecies *cathpharius* and possibly *ludlowi*): oak, rhododendron and suchlike broad-leaved forest in moist-deciduous and evergreen biotope. General habits typically woodpecker. Actions and behaviour—flying from one dead tree-trunk to another and searching each methodically—reminiscent of Tree Creeper. Food: insects and grubs; also some flower nectar (rhododendron). Call: an occasional soft, clear *pwik*; a loud monotonous *chip* at intervals while searching a tree, more quickly uttered in flight.

160 FULVOUSBREASTED PIED WOODPECKER

Picoides macei (Vieillot)

Size Bulbul — ; length 19 cm (7½ in.).

Field Characters A smallish pied woodpecker with white-barred black back. Crown and crest crimson in male, black in female; a black moustachial streak continued back to behind ear-coverts. *Below*, throat and foreneck uniform pale rufous-brown; breast fulvous, faintly black-streaked; abdomen and vent barred blackish and grey; under tail-coverts crimson.

Status, Habitat, etc. Resident. Duars and foothills up to 2500 m—open forest and wooded country. Typical woodpecker. Commonly met in pairs or family parties of 3 or 4, often working on the same tree directly upward or in spirals. Food: insects and grubs; also some seeds and berries. Call: a moderately loud *pik ... pik* at intervals while feeding; a shrill, rapid *pik-pipipipipipipipipipi* when excited.

161 GREYCROWNED PIGMY WOODPECKER

Picoides canicapillus (Blyth)

Size Sparrow — ; length 14 cm (5½ in.).

Field Characters A diminutive pied woodpecker. Male. *Above*, forehead and crown ashy grey; short occipital crest scarlet, surrounded by black. A broad whitish supercilium from behind eye continued as a wide band down sides of neck. Upper back black; wings, lower back and rump black broadly barred with white. Upper tail-coverts and tail black, the lateral rectrices more or less barred with white. *Below*, chin and throat whitish streaked with ashy; rest fulvous-brown streaked with black. Female similar but scarlet of hindcrown replaced by black.

Status, Habitat, etc. Resident. Duars, foothills and up to 1700 m, occasionally to 2000 m—open oak forest and mixed secondary jungle. Pairs, commonly in association with itinerant hunting parties of tits, nuthatches, etc. Creeps actively along and around end twigs, like a nuthatch, tapping energetically with bill to dislodge insects. Flight rather sparrow-like, without the characteristic woodpecker bounds. Food: ants, caterpillars, etc.; also pulp of soft fruits and some flower nectar. Call: an occasional feeble mousy *click-r-r* (contact call?) while hunting; a soft but far-carrying mechanical drumming on a thin branch.

162 REDEARED BAY WOODPECKER

Blythipicus pyrrhotis (Hodgson)   p. 48

Size Pigeon − ; length 25 cm (10 in.).

Field Characters A medium-sized barred rufous-and-black woodpecker with greenish yellow bill. Male has a prominent scarlet cross-band on nape extending to behind ear-coverts. Female similar, but without the scarlet nape-band. Casually confusable with Rufous Woodpecker (150), but larger size, greenish yellow bill, and evergreen forest habitat diagnostic.

Status, Habitat, etc. Resident. Duars, foothills and up to 2000 m—dense evergreen climax forest and adjacent secondary jungle with bamboo. Usually pairs, rarely 3 or 4 together. Keeps to lower storey in dense forest, avoiding opener parts. Feeds chiefly among roots and at base of decaying stumps, rarely high up. Food: mainly termites and beetle larvae. Call: a distinctive, unmusical but not unpleasant *chake, chake* ... uttered slowly and deliberately 4 or 5 times, getting faster and fainter at the finish. A loud chattering *kererē-kererē-kererē* when agitated, accompanied by excited wing-flicking.

163 LARGER GOLDENBACKED WOODPECKER

Chrysocolaptes lucidus (Scopoli)   p. 45

Size Pigeon; length 35 cm (14 in.).

Field Characters A medium-sized golden-backed woodpecker. Male. *Above*, crown and crest crimson; nape and hindneck white. Back and mantle golden olive; rump crimson; upper tail-coverts and tail black. *Below*, buffy white: foreneck and breast black, spotted with white. Cheeks and chin white; two black lines on each cheek meeting to continue as one down sides of throat; a fifth broader stripe down centre of chin and foreneck. Female similar, but crimson of the head replaced by black stippled with white. Casually confusable with Golden-backed 3-toed Woodpecker (154) q.v. but pattern of throat-stripes distinctive.

Status, Habitat, etc. Resident. Duars and foothills chiefly below 700 m; sparingly up to 1600 m (subspecies *guttacristatus*): moist-deciduous and semi-evergreen biotope, chiefly well-wooded broken hummocky country. Typical woodpecker in habits, food and behaviour, with the characteristic bounding flight. Call: a shrill tinny discordant 'laugh' uttered chiefly on the wing. Also the usual mechanical drumming on wood.

BROADBILLS: Eurylaimidae

164 COLLARED BROADBILL *Serilophus lunatus* (Gould)   p. 49
Size Bulbul ± ; length 19 cm (7½ in.).
Field Characters A sluggish ashy grey arboreal bird with a short recumbent crest overhanging the nape. Lower back chestnut. Wings black, with contrasting chestnut, blue and white markings. Tail black, graduated, the lateral rectrices tipped with white. Prominent yellow skin around eye. Female similar, but with a demi-gorget of white-tipped feathers on each side of neck.
Status, Habitat, etc. Resident. Duars, foothills and up to 1700 m in mixed secondary tree and bamboo jungle—tropical semi-evergreen and evergreen biotope (subspecies *rubropygius*). Loose parties of 5 to 20; active early morning and around dusk, rather lethargic during daytime. Plucks insects off branches and foliage in clumsy fluttering sallies. Volplanes, or bounces lightly, from branch to branch in the quest in a manner reminiscent of Wood Shrike. Food: grasshoppers, mantises and other insects. Call: a soft musical whistle; a low *chir-r-r* uttered at rest and on the wing; a low mouse-like squeaking when alarmed.

165 LONGTAILED BROADBILL
   Psarisomus dalhousiae (Jameson)   p. 49
Size Bulbul + ; length 25 cm (10 in.).
Field Characters A showy grass-green arboreal bird with a longish graduated tail, bright blue above, black below. Crown black with an oblong yellow spot on each side and a bright blue patch in the centre. Chin and throat bright yellow, this colour projected up sides of neck as a collar, broken behind. A white patch on black underside of wings conspicuous in flight. Sexes alike.
Status, Habitat, etc. Resident, with seasonal local movements. Duars, foothills and up to 2000 m, in tropical and subtropical evergreen biotope—mixed secondary forest and bamboo, riverain jungle, etc. Arboreal and sociable. Flocks of 15 to 30 keeping to the foliage canopy or middle storey vegetation. Actions and behaviour similar to Collared Broadbill's (164) q.v.; reminiscent also of minivets. Food: large insects and spiders. Call: a distinctive loud, sharp whistle *tseeay*,

tseeay ... repeated 5 to 8 times on an unvaried pitch, usually in flight; an occasional single note *sĕweet* while foraging.

PITTAS: Pittidae

166 **BLUENAPED PITTA** *Pitta nipalensis* (Hodgson)   p. 49
Size Quail + ; length 25 cm (10 in.).
Field Characters A dumpy stub-tailed thrush-like terrestrial bird. Male largely blue and greenish brown above, plain fulvous below. Nape and hindneck bright blue; tail brown tinged with green. Female. *Above*, similar to male but hindcrown rufous-fulvous instead of blue, and hindneck green. *Below*, throat usually more whitish. Flight silhouette from side reminiscent of waterhen (*Amaurornis*).
Status, Habitat, etc. Resident, subject to seasonal movements largely altitudinal. Duars, foothills and up to 2000 m—tropical and subtropical secondary evergreen and bamboo jungle, densely overgrown clearings, etc. Predominantly terrestrial, shy and skulking. Singly or separated pairs in dense shrubbery flicking aside the fallen leaves and digging into the mulch for food. Progresses in swift long hops. Food: ants, beetles and other insects, and other creeping small animals. Call: a beautiful double whistle, heard mostly early mornings and at dusk.

167 **HOODED or GREENBREASTED PITTA**
   Pitta sordida (P. L. S. Müller)   p. 45
Size Quail; length 19 cm (7½ in.).
Field Characters A multicoloured, largely green, dumpy, stub-tailed terrestrial bird. *Above*, crown and nape rich rufous-brown; cheeks, ear-coverts and throat black, continued as a collar on hindneck. Mantle, back and rump bluish green. Upper tail-coverts and a shoulder-patch bright blue. Wing-quills black with a prominent white patch; tail black tipped with blue. *Below*, breast and sides of body pale greenish blue; belly black; vent and under tail-coverts bright crimson. Sexes alike.
Status, Habitat, etc. Resident, with some altitudinal and dispersal movements. Duars, foothills, and up to 2000 m—subtropical moist-deciduous and evergreen biotope: secondary forest and scrub jungle (subspecies *cucullata*). Mainly solitary and terrestrial. Hops on ground digging in the mulch for food. Flies up into low branches when disturbed, sitting motionless, only the stub tail wagging slowly up and down. Food: insects and worms. Call: a loud musical whistle.

168 **BLUE PITTA** *Pitta cyanea* Blyth
Size Quail + ; length 23 cm (9 in.).

Field Characters A brightly coloured pitta, largely blue. **Male.** Upperparts and tail blue. Forehead and crown greenish grey, with a black median streak; a short scarlet nuchal crest; a broad black stripe from bill through eye to nape; black moustachial streaks. A white patch in wing, concealed at rest, prominent in flight. *Below*, throat white; rest pale blue (yellowish on breast) spotted and brokenly barred black. **Female** has upperparts dull brownish suffused with pale blue; rump, upper tail-coverts and tail pure blue.

Status, Habitat, etc. Resident, subject to local migrations. Not recorded from Sikkim. Evergreen duars and foothills, rarely up to 2000 m: dank ravines and scrubby undergrowth in mixed tree and bamboo forest. Habits and food as in Hooded Pitta (167), q.v. Call: a clear, full, double whistle.

LARKS: Alaudidae

169 LONGBILLED CALANDRA LARK
Melanocorypha maxima Blyth

Size Myna − ; length 21 cm (8 in.).

Field Characters A large, dumpy thick-billed lark with much white in tail and a prominent black spot on each side of breast. *Above*, brown, tinged with rufous on head and rump, the feathers dark-centred; supercilium and cheeks whitish. *Below*, whitish. Sexes alike.

Status, Habitat, etc. Resident in the Tibetan Plateau facies above 3600 m—barren steppe country, and humpy bogs and marshland around high-elevation lakes (subspecies *maxima*). Pairs and small scattered parties feeding on ground. Food: seeds and insects. Call: low musical whistles when disturbed and flying off. Song, a series of rather feeble disconnected strophes often interlarded with mimicry of other birds' calls.

170 HORNED LARK *Eremophila alpestris* (Linnaeus)

Size Bulbul; length 20 cm (8 in.).

Field Characters A largish high-elevation lark, chiefly pinkish ashy brown above, white below. Face and throat whitish with contrasty black cheeks separated from a broad black gorget across upper breast by a white band. A black band across crown with a horn-like tuft of feathers above each eye. **Female** has the crown ashy brown, lightly streaked with black, without the cross-band or horns; black of cheeks and breast more restricted.

Status, Habitat, etc. Resident between 3600 m and 5000 m, descending lower in winter (subspecies *elwesi*): bleak stony hillsides in typical Tibetan steppe country. Pairs and small scattered parties; sometimes fairly large flocks in non-breeding season. Runs about in short spurts like a small plover or wagtail. Flies with deliberate leisurely-looking wing beats. Horns raised under excitement, e.g. when facing a rival or in courtship. Food: weed seeds and insects. Call: an occasional single, rather mournful *peo* uttered in flight. Song, of squeaky disjointed strophes, feeble in volume, recalling a leaf warbler's.

171 **SMALL SKYLARK** *Alauda gulgula* Franklin   p. 52
Size Sparrow; length 15 cm (6 in.).
Field Characters A dark-streaked sandy brown lark, with white outer tail-feathers. Mistakable for a pipit, but squatter in build and with relatively short tail. An indistinct tuft-like crest, noticeable when partially raised. Underparts pale fulvous-buff, darker on breast and faintly streaked and spotted with blackish. Sexes alike.
Status, Habitat, etc. Apparently a winter visitor to C. and E. Bhutan and Arunachal Pradesh (above 1600 m) from southern Tibet and Yunnan (subspecies *vernayi*). Small scattered parties on open grassy hilltops and plateaus, running about in short spurts picking weed seeds etc. Food: seeds, tiny bulbils and insects. Call: unrecorded specifically; presumably as in other races including the lively, sustained aerial song which is the skylark's crowning distinction.

SWALLOWS, MARTINS: Hirundinidae

172 **SWALLOW** *Hirundo rustica* Linnaeus (2 subspecies)   p. 52
Size Sparrow $\pm$, with long forked tail; total length 18 cm (7 in.).
Field Characters *Above*, glossy steel-blue. *Below*, pale pinkish white in subspecies *gutturalis*, chestnut in *tytleri* (illustrated). Forehead, chin and throat chestnut, the last bordered below by a broken blue-black breast-band in *gutturalis*; pectoral collar confined to sides of upper breast in *tytleri*. Tail deeply forked, all feathers except middle pair with a white spot, visible when bird banks in flight. Sexes alike.
Status, Habitat, etc. *Gutturalis* summer (breeding) visitor from peninsular India; *tytleri* winter visitor from NE. Asia and Kamchatka. Duars, foothills and up to 3000 m seasonally; open country, cultivation, neighbourhood of jheels, and shingly river-beds. Hawks winged insects usually low over meadows, standing crops, reed-beds, etc. Highly gregarious in winter. Flight swift: a few rapid wing beats followed by a swooping glide with agile banking and turning movements. Roosts

communally in vast swarms in reed-beds or shrubbery near water.
Food: midges, gnats, flies, etc. Call: pleasant twittering notes uttered
in flight or at perch. Song of male, a feeble but lively musical twittering,
continuously for several seconds.

173 **WIRETAILED SWALLOW** *Hirundo smithii* Leach
Size Sparrow ± ; length 14 cm (5½ in.) excluding tail-wires.
Field Characters Glossy steel-blue above, with a bright chestnut cap.
Readily distinguished from other swallows by glistening pure white
underparts, and two long fine 'wires' projecting from tail. Sexes alike,
but tail-wires of female shorter.
Status, Habitat, etc. Resident, subject to summer-winter altitudinal
movements. Duars, and foothills, normally up to 1500 m, rarely 2500 m:
neighbourhood of cultivation and water—rivers, jheels, reservoirs, etc.
Loose parties hawking tiny winged insects near the ground or water
surface. Habits, behaviour and food similar to 172, q.v. Call: a low
chit-chit in flight. Song of male, a musical twittering double *chirrik-
weet*, *chirrik-weet* repeated intermittently from perch near nest.

174 **REDRUMPED SWALLOW** *Hirundo daurica* Linnaeus   p. 52
Size Sparrow ± ; length 14 cm (5½ in.) excluding long deeply forked tail.
Field Characters *Above*, glistening steely blue-black; supercilium and
sides of head chestnut connected by a chestnut half-collar on hindneck.
Rump pale chestnut (almost whitish), diagnostic in flight. *Below*,
fulvous white coarsely streaked with dark brown (subspecies *nipalensis*).
Sexes alike.
Status, Habitat, etc. Summer (breeding) visitor between 1000 and 2500 m;
descending to lower levels and the Indian plains in winter. Affects the
neighbourhood of upland pastures and hill villages. Habits, behaviour
and food similar to 172, q.v. Call: a loud low-pitched *cheer* and a
sparrow-like chirp, uttered in flight. Song of male, a cheerful subdued
continuous twittering.

175 **HOUSE MARTIN** *Delichon urbica* (Linnaeus)   p. 37
Size Sparrow; length 15 cm (6 in.).
Field Characters A black and white swallow with short forked tail. *Above*,
glossy blue-black with smoky white rump. *Below*, smoky white including
feathered ('stockinged') legs and feet. Sexes alike. Casually confusable
in flight with House Swift which has white rump; but white underparts
(*v.* black in swift) diagnostic (subspecies *cashmeriensis*). Sexes alike.
Status, Habitat, etc. Resident and partial local and altitudinal migrant.
Breeds between 1500 and 5000 m; winters down to the duars and plains.

Plate 20, artist Robert Scholz

WREN-BABBLERS, GROUND, BUSH AND LEAF WARBLERS, GOLDCREST, TIT

1 R SPOTTED SHORT-TAILED WREN-BABBLER,
Spelaeornis formosus   page 132
    Sparrow − . 1200–2500 m.

2 R TAILED WREN-BABBLER, *Spelaeornis caudatus*   131
    Sparrow − . Ferruginous throat extends to breast
    and flanks. 1800–2400 m.

3 R CHESTNUTHEADED GROUND WARBLER, *Tesia*
castaneocoronata   172
    Sparrow − . Diminutive. Stub tail. Bright yellow
    throat. Duars to 3300 m.

4 R BROWN WREN-BABBLER, *Pnoepyga pusilla*   131
    Sparrow − . Diminutive. 1500–3000 m.

5 R GOLDCREST, *Regulus regulus*   189
    Sparrow − . Diminutive. 2 yellowish wing-bars.
    Crown with orange median and black lateral bands.
    2200–4000 m.

6 R FIRECAPPED TIT, *Cephalopyrus flammiceps*   221
    Sparrow − . Diminutive. ♀ Single wing-bar. Yellow
    forehead, yellowish rump. Underparts yellowish olive
    and buff. 300–2300 m.

7 R SPOTTED BUSH WARBLER, *Bradypterus thoracicus*   175
    Sparrow − . Duars to 4300 m.

8 M DUSKY LEAF WARBLER, *Phylloscopus fuscatus*   183
    Sparrow − . No wing-bar. Greyish supercilium.
    Duars

9 R ORANGEBARRED LEAF WARBLER, *Phylloscopus*
pulcher   183
    Sparrow − . 2 orange-yellow wing-bars. Long yellow
    supercilium. Faint coronal median stripe. 500–
    4300 m.

10 R GREENISH LEAF WARBLER, *Phylloscopus*
trochiloides   185
    Sparrow − . Single faint wing-bar. Yellow super-
    cilium. Dark eye-streak. Duars to 3700 m.

R. Scholz

Inches
0   2   4

D.M.H.
1956

Plate 21, artist D. M. Henry

BABBLERS, SHRIKE-BABBLER, TIT-BABBLERS, MYZORNIS, YUHINAS

1 R GOLDENHEADED BABBLER, *Stachyris chrysaea* page 133
Sparrow – . Crown and nape black-streaked. Short black moustachial stripe. 1200–2600 m.

2 R FIRETAILED MYZORNIS, *Myzornis pyrrhoura* 148
Sparrow – . Red-and-green tail. ♀ duller. 1600–3600 m.

3 R CHESTNUT-THROATED SHRIKE-BABBLER,
Pteruthius melanotis 150
Sparrow – . ♀ Throat buffish. Cinnamon moustache. Wing-bars pale rufous. 800–2700 m.

4 R STRIPETHROATED YUHINA, *Yuhina gularis* 154
Sparrow. Erectile brown crest. Orangey stripe on wings. Foothills to 3600 m.

5 R GOLDENBREASTED TIT-BABBLER, *Alcippe chrysotis* 155
Sparrow – . Silver-grey ear-coverts. Orange wing-patch. 2400–3000 m.

6 R BLACKTHROATED BABBLER, *Stachyris nigriceps* 134
Sparrow – . Crown blackish, boldly white-striped. Black supercilium to nape. 750–1800 m.

7 R REDFRONTED BABBLER, *Stachyris rufifrons* 132
Sparrow – . White chin. Faintly streaked ochraceous throat. Whitish belly. Up to 900 m.

8 R QUAKER BABBLER, *Alcippe nipalensis* 157
Sparrow – . White eye-ring. Blackish supercilium to nape. Lower foothills to 2400 m.

9 R YELLOWNAPED YUHINA, *Yuhina flavicollis* 153
Sparrow – . Erectile chocolate-brown crest. White eye-ring. Duars to 3000 m.

10 R WHITEBROWED YUHINA, *Yuhina castaniceps* 153
Sparrow – . Crown grey, scalloped paler. Rufous-brown ear-coverts. Rectrices white-tipped. 600–1500 m.

11 R CHESTNUTHEADED TIT-BABBLER, *Alcippe castaneceps* 156
Sparrow – . Crown white-streaked. White supercilium and ear-coverts. Brown stripe behind eye. Black and rufous shoulder-patch. 1500–3000 m.

12 R WHITEBROWED TIT-BABBLER, *Alcippe vinipectus* 156
Sparrow – . Head dark brown. White supercilium to nape. 1500–4200 m.

Affects open valleys, grassy hillsides with cliffs, and environs of mountain villages and terraced cultivation. Highly gregarious even when nesting. A typical swallow in habits, behaviour and food, cf. 172. Very agile on the wing. Call: low squeaks and twittering.

176 **NEPAL HOUSE MARTIN** *Delichon nipalensis* Moore   p. 37
Size Sparrow ± ; length 13 cm (5 in.).
Field Characters A small black and white swallow, very similar to 175, but with short square-cut tail (*v.* forked), blackish throat and glossy black under tail-coverts (*v.* entirely glistening white underparts).
Status, Habitat, etc. Resident, with some local and altitudinal movements. Breeds colonially between 2000 and 4000 m; descends to 350 m in the foothills in winter: river valleys, mountain ridges with vertical cliffs, etc. Highly sociable, and very similar in all respects to the foregoing species (175).

SHRIKES: Laniidae

177 **TIBETAN GREYBACKED SHRIKE**
       Lanius tephronotus (Vigors)   p. 64
Size Bulbul + ; length 25 cm (10 in.).
Field Characters A high-elevation shrike with the characteristic stout hooked bill. *Above*, narrow forehead and a broad band through eyes to ear-coverts black. Crown, nape, hindneck and back dark leaden grey; rump and upper tail-coverts rufous; tail chestnut-brown. Wings black, sometimes with an insignificant white patch or 'mirror'. *Below*, rufous, paler on belly. Sexes alike.
Status, Habitat, etc. Breeds between 2700 and 4500 m in summer; winters at lower elevations, and in the foothills and adjacent plains: open scrub and secondary growth in and around hill cultivation and clearings. Territorial and parochial; singly or widely separated pairs. Perches upright on exposed bushtops etc. keeping a sharp look-out for crawling prey, pouncing on it swiftly and silently. Sometimes takes flying insects by aerial sallies. Excess food is impaled on thorns by way of a larder. Food: large insects, lizards, frogs, young or sickly birds and rodents, etc. Call: harsh scolding cries reminiscent of squealing of frog caught by a snake. Song (in breeding season), a beautiful subdued musical soliloquy, sustained for several minutes, interlarded with expertly mimicked calls of other birds.

178 **BLACKHEADED SHRIKE** *Lanius schach tricolor* (Hodgson) p. 52
Size Bulbul + ; length 25 cm (10 in.).

Field Characters *Above*, crown, sides of head, nape and hindneck black. Upper back ashy grey, changing to light chestnut on rest of upper plumage. Wings blackish, with a small white 'mirror' very conspicuous in flight. Tail black, graduated, tipped and margined rufous. *Below*, cheeks, chin and throat white shading to pale rufous on breast and to bright rufous on rest of lower plumage except belly which is whitish. Sexes alike.

Status, Habitat, etc. Resident. Breeds between 3000 and 4000 m in summer; winters at lower elevations, and in the foothills and adjacent plains. Habits, behaviour, food and calls very similar to Greybacked Shrike (177), q.v.

179 **BROWN SHRIKE** *Lanius cristatus* Linnaeus p. 52
Size Bulbul ± ; length 19 cm (7½ in.).

Field Characters *Above*, forehead and supercilium white; a black stripe through eye to ear-coverts; crown, mantle, rump and upper tail-coverts reddish brown. Wings brown without white 'mirror'; tail rufous-brown. *Below*, chin, cheeks and throat white; rest pale fulvous variably tinged with rufescent on breast and belly. Sexes alike. Some adults have breast and flanks variably spangled with dark crescentic bars.

Status, Habitat, etc. Winter visitor (September to March/April): duars, foothills, and up to 1800 m. Commoner at low and moderate elevations—forest edges and clearings, open secondary jungle, grassy hillsides with scattered bushes, etc. Habits, behaviour and food typical of the family, cf. 177. More crepuscular than the other two; commonly seen hunting till past dusk when also it is particularly noisy. Call: a harsh, loud, swearing *chr-r-r*. Only an occasional subdued warbling sub-song while in winter quarters.

ORIOLES: Oriolidae

180 **GOLDEN ORIOLE** *Oriolus oriolus* (Linnaeus) p. 52
Size Myna; length 25 cm (10 in.).

Field Characters Male, bright golden yellow with black in wings and tail. A conspicuous black streak through eye. Female and young male duller and greener yellow with whitish underparts streaked with brown. Some adult females golden like adult male (subspecies *kundoo*).

Status, Habitat, etc. Rare straggler to Jalpaiguri duars. Singly or pairs: well wooded country, village groves, orchards, etc. Food: insects and berries. Call: a harsh *chee-ah*; also rich fluty melodious notes *pee-lo* or *pee-lo-lo*.

181 **BLACKNAPED ORIOLE** *Oriolus chinensis* Linnaeus p. 52
Size Myna; length 25 cm (10 in.).
Field Characters M a l e, overall bright golden yellow with black-patterned wings and tail. A black band through eye continued behind to meet over nape. F e m a l e duller and greener (subspecies *tenuirostris*).
Status, Habitat, etc. Rare winter visitor (or resident ?); recorded only October to March. Duars, foothills, and up to 2000 m in well-wooded country. Seen singly as a rule. Chiefly arboreal, keeping to leafy fruit-bearing trees. F o o d: fruits, berries and insects. C a l l: melodious, liquid, flute-like; also a harsh, oft-repeated grating note.

182 **BLACKHEADED ORIOLE** *Oriolus xanthornus* (Linnaeus)
Size Myna; length 25 cm (10 in.).
Field Characters Brilliant golden yellow with jet black head, throat and upper breast, and black patterning in wings and tail. Crimson eyes and flesh-pink bill conspicuous. Sexes alike, but upperparts of female tinged with olive. Young birds (both sexes) have yellow forehead, blackish crown and sides of head streaked with yellow; chin and throat white, breast yellow streaked with black.
Status, Habitat, etc. Resident, subject to local movements. Duars and foothills, mostly under 1000 m, in well-wooded country—open forest, groves of trees around villages, etc. Singly or pairs, often in loose association with the itinerant mixed hunting parties. Arboreal and largely frugivorous. Flight strong and dipping. F o o d: fruits—chiefly wild figs—berries, and insects. C a l l: a loud harsh *cheeah* or *kwaak* at intervals; also melodious fluty whistles *wye-you* or *wye-you-you*.

183 **MAROON ORIOLE** *Oriolus traillii* (Vigors) p. 64
Size Myna + ; length 30 cm (12 in.).
Field Characters M a l e, overall glossy crimson-maroon. Head, neck and wings black; tail chestnut-maroon. F e m a l e (and young male) similar, but underparts greyish white streaked with blackish. In overhead flight square-cut claret-crimson tail diagnostic.
Status, Habitat, etc. Resident, subject to seasonal local movements. Duars, foothills and up to 2000 m, usually below 1000—moist-deciduous and evergreen forest. Singly or pairs. Keeps to tall trees, commonly in association with drongos, minivets, etc. F o o d: wild figs and berries, insects, flower nectar. C a l l: a harsh *cheeah* and rich fluty liquid whistles *pi-lo-lo*; similar in pattern to Blackheaded Oriole's.

DRONGOS: Dicruridae

184 **BLACK DRONGO** *Dicrurus adsimilis* (Bechstein) p. 53
Size Bulbul + ; length, including long forked tail, 30 cm (12 in.).
Field Characters A slim glossy jet black bird with long, deeply forked tail.
Sexes alike.
Status, Habitat, etc. Resident, subject to seasonal altitudinal movements.
Duars, foothills and up to 2000 m; at the lower elevations in winter
(subspecies *albirictus*). Affects light forests of oak, rhododendron, etc.,
and openly wooded country on outskirts of cultivation and habitations.
Singly or scattered pairs, perched on tree tops and similar look-out
posts. Takes flying insects on the wing, like a flycatcher, or pounces on
them on the ground. Bold and pugnacious; very swift and agile on the
wing. Often pirates food from larger birds, swooping on them with
audacity and forcing them to yield their lawful prize. F o o d : grasshoppers, moths and other insects; occasionally lizards and weakling birds;
also flower nectar. C a l l : a defiant-sounding harsh double *ti-tiu* like a
shikra's; also a harsh *cheece-cheece-chichuk*. Some pleasant notes in
breeding season.

185 **GREY DRONGO** *Dicrurus leucophaeus* Vieillot p. 53
Size As of Black Drongo, Bulbul + ; overall length 30 cm (12 in.).
Field Characters A slim glossy slate-black drongo with long deeply
forked tail and conspicuous crimson eyes. Underparts dull slaty.
Sexes alike. Distinguished from Black Drongo by slimmer build,
longer deeper-forked tail, dull grey underparts (*v.* glossy black) and
ruby-red eyes.
Status, Habitat, etc. Breeds in the foothills and up to 3000 m in summer;
descends lower and into the Indian plains in winter (September to
March/April) (subspecies *longicaudatus* and *hopwoodi*). Pine, oak and
rhododendron forest in summer; better wooded areas than Black
Drongo in winter. Habits, behaviour and food as in Black Drongo, q.v.;
particularly fond of flower nectar. C a l l : a variety of harsh screeches
and pleasant musical whistling notes, with perfectly mimicked calls of
other birds freely interlarded.

186 **CROWBILLED DRONGO** *Dicrurus annectans* (Hodgson)
Size Myna + ; length 25 cm (10 in.) to tip of tail.
Field Characters A glossy jet black drongo, very like the Black but with
larger, heavier bill and shorter, less deeply forked tail. Distinguished
also by stockier build and forest habitat (*v.* open cultivation). Sexes
alike.

Status, Habitat, etc.   Resident (or breeding summer visitor ?). Seasonal
movements unclear. Duars and lower foothills rarely above 700 m:
densely forested country—moist-deciduous and evergreen. Habits and
behaviour largely as of Black Drongo, but little specifically known.
Food: crickets, grasshoppers, termites and other insects; doubtless
also other small animals. Call: similar to Black Drongo's, clear and
loud and pleasanter.

187 **BRONZED DRONGO** *Dicrurus aeneus* Vieillot   p. 53
Size Bulbul + ; length, including longish forked tail, 24 cm (9½ in.).
Field Characters A small black forest drongo, highly glossed with metallic
bronze-green and blue. Bill broad and flattened like a flycatcher's. In
dim light superficially very like Black Drongo but much smaller and with
less deeply forked tail. Sexes alike.
Status, Habitat, etc.   Resident, subject to seasonal altitudinal movements.
Duars, foothills and up to 2000 m: moist-deciduous and evergreen
forest biotope. Prefers broken foothills country—glades, forest edges
and mixed bamboo jungle. Territorial and parochial, frequenting same
spot day after day. Makes sprightly aerial sallies to snatch winged
insects, and loops gracefully back to the same perch. Food: insects—
midges, flies, winged ants, etc.; also flower nectar. Call: a wide range
of loud clear musical whistles, interwoven with perfect mimicry of
other birds' calls.

188 **LESSER RACKET-TAILED DRONGO**
   Dicrurus remifer (Temminck)   p. 53
Size Myna + ; length 30 cm (12 in.) plus 9 cm to end of tail.
Field Characters A glossy metallic black forest drongo with a velvety
pad-like tuft of feathers above base of bill. Two elongated wire-like
outer tail-feathers ending in 'rackets' (webbed on both sides of shaft
contra 190, q.v.). Sexes alike. In the flying bird these rackets look like
a pair of bumble-bees in hot pursuit!
Status, Habitat, etc.   Resident. Duars, foothills and up to 2000 m: heavy
evergreen and moist-deciduous forest. Entirely arboreal. Singly or in
twos and threes, often in company with the roving hunting parties.
Keeps to dense canopy at edge of glades, firelines, forest streams, etc.
Hawks winged insects by agile looping aerial sallies in flycatcher style.
Flight dipping and noisy. Food: chiefly insects; also flower nectar.
Call: a wide range of loud metallic whistles interlarded with excellently
mimicked calls of other birds. Indistinguishable from notes of Large
Racket-tailed (190).

189 HAIRCRESTED or SPANGLED DRONGO

Dicrurus hottentottus (Linnaeus)   p. 53

Size Myna ± with longer tail; overall length 30 cm (12 in.).

Field Characters An iridescent, glossed and spangled blue-black drongo with distinctive longish, almost square-cut tail, the longest outermost feathers curled at the ends. A few recumbent hair-like plumes extending backward from forehead over crown (visible only in profile at short range). Bill long and noticeably curved. Sexes alike.

Status, Habitat, etc. Resident, subject to seasonal movements and wandering. Duars, foothills—usually at low levels, sometimes up to 2000 m: moist-deciduous and evergreen biotope. Partial to coppices of flowering trees such as Silk Cotton (*Bombax*) and Coral (*Erythrina*). Predominantly a nectar-feeder with specially adapted bill and tongue. Flight noisy, with thudding wing beats. Food: flower nectar and insects. Call: a subdued metallic clanging note, constantly uttered; loud clear whistles mixed with mimicry of other birds' calls. Very noisy in breeding season.

190 LARGE RACKET-TAILED DRONGO

Dicrurus paradiseus (Linnaeus)   p. 53

Size Myna; with tail 35 cm (14 in.) long.

Field Characters A large metallic black drongo with a conspicuous backwardly curving crest from forehead, and two long spatula-tipped wire-like streamers in tail. In flight these 'rackets' trail behind looking very like a pair of angry bumble-bees chasing the bird! Sexes alike.

Status, Habitat, etc. Resident. Duars, foothills—usually at low elevations; sometimes up to 1500 m: deciduous and evergreen biotope in well-wooded broken country—sal, teak and bamboo jungle, tea gardens, etc. (subspecies *grandis*). Habits and behaviour typical of drongos—similar to Lesser Racket-tailed and Haircrested, qq.v. Flight rather noisy and dipping—rapid wing beats punctuated with short pauses. Food: chiefly large insects; also lizards etc. and flower nectar. Call: a wide range of loud metallic, melodious notes and whistling, varied with perfect mimicry of other birds' calls.

SWALLOW-SHRIKES: Artamidae

191 ASHY SWALLOW-SHRIKE *Artamus fuscus* Vieillot   p. 52

Size Bulbul ± ; dumpier, with short tail. Length 19 cm (7½ in.).

Field Characters A dark slaty grey short-tailed bird, with paler grey rump and underparts. Tail slaty black, tipped with white. Bill bluish, heavy, finch-like. Sexes alike. At rest tips of closed wings reach end of tail.

Status, Habitat, etc.   Resident and marked local migrant. Low elevations, locally up to 1700 m, in dry- and moist-deciduous openly wooded country, especially where abounding in palms. Parties or small flocks. Perches in exposed situations, e.g. telegraph wire, all huddled together, bobbing the stumpy tail slowly and screwing it from side to side. Makes aerial sallies to capture winged insects and sails back gracefully to the perch. Flight: several rapid wing beats followed by an effortless glide, reminiscent of bee-eater. Food: butterflies, dragonflies and other insects. Call: a distinctive, harsh *chēk-chēk-chēk* uttered at rest and in flight.

MYNAS, STARLINGS: Sturnidae

192 SPOTTEDWINGED STARE *Saroglossa spiloptera* (Vigors)   p. 64

Size Bulbul ± ; length 19 cm (7½ in.).

Field Characters A small trim myna reminiscent of the Greyheaded. Male. *Above*, brownish grey scalloped with black. A prominent white patch on the blackish wings. *Below*, chin and throat deep chestnut; rest rusty white. Female. *Above*, sooty brown. White wing-patch as in male. *Below*, pale fulvous ashy brown, scalloped with white, especially on chin and throat. Pointed wings and typical starling-like flight, together with dark head, chestnut rump and white wing-patch (as in Blueheaded Rock Thrush) diagnostic.

Status, Habitat, etc.   Status unestablished; apparently east-west migrant along Himalayan foothills between 700 and 1200 m: March/April and June/July. Affects open forest. Arboreal and gregarious. Sometimes large flocks feeding on flowering trees in company with other mynas. Food: flower nectar, fruits and insects. Call: a noisy chatter like that of Greyheaded and Jungle mynas when in a flock; a soft chirruping *chik-chik* when sitting individually.

193 GREYHEADED MYNA *Sturnus malabaricus* (Gmelin)

Size Myna − ; length 21 cm (8½ in.).

Field Characters A small, trim, silvery grey-and-rufous myna. *Above*, head and upperparts silvery or brownish grey. Tail largely rufous; wing-quills black and grey. *Below*, chin, throat and breast lilac-grey or pale rufous; rest bright rufous. Sexes more or less alike.

Status, Habitat, etc.   Resident, subject to marked seasonal local migrations and nomadism. Duars and foothills, normally up to 1500 m: thinly wooded country, open secondary jungle, forest plantations, etc.

(subspecies *malabaricus*). Gregarious, arboreal and frugivorous. Flocks, often large, associated with other mynas on flowering and fruiting trees. Flight swift and typical: a few rapid wing beats followed by a short glide. Food: fruits and berries, flower nectar, insects. Call: sharp disyllabic metallic notes on the wing; a pleasant tremulous single whistle or warble at rest.

194 PIED MYNA *Sturnus contra* Linnaeus

Size Myna ± ; length 23 cm (9 in.).

Field Characters A trim black-and-white myna with conspicuous deep orange-red naked patch round eye and shining whitish ear-coverts; bright orange and yellow bill. Sexes alike.

Status, Habitat, etc. Resident, with some seasonal movements. Duars and lower foothills up to 600 m (subspecies *contra* and *sordidus*): well-watered cultivated country, and around habitations—cattle pens, garbage dumps and damp grazing grounds, etc. Sociable, predominantly insectivorous, and ground-feeding; noisy squabbling parties frequently with cattle. Food: insects, fruits and cereal grains. Call: high-pitched pleasant musical whistling notes.

195 INDIAN MYNA *Acridotheres tristis* (Linnaeus)

Size Pigeon − ; length 23 cm (9 in.).

Field Characters A perky, well-groomed dark brown bird with glossy black head and white-tipped tail. A large white wing-patch conspicuous in flight. A naked patch behind eye, bill and legs bright yellow. Sexes alike.

Status, Habitat, etc. Resident, with some summer-winter altitudinal movements. Duars and lower foothills; but breeding locally sometimes up to 3000 m: neighbourhood of habitations and cultivation. Pairs and family parties while foraging; enormous noisy gatherings at communal roosts in large trees, reed-beds, etc. Food: all-embracing: fruits, insects, kitchen scraps, cereal grains, flower nectar, or whatever. Call: a variety of high-pitched falsetto notes and chatter; a loud scolding *rādio, rādio . . .* ; an amusing *keek-keek-keek, kok-kok-kok, churr-churr*, etc. accompanied by ludicrous bobbing of head with plumage frowzled by male before his mate.

196 BANK MYNA *Acridotheres ginginianus* (Latham)

Size Myna − ; length 21 cm (8½ in.).

Field Characters Like Indian Myna but slightly smaller and pale bluish grey overall instead of vinous-brown. Wing-patch and tips of tail-feathers pinkish buff instead of pure white. Naked skin behind eye brick-red instead of yellow. Sexes alike.

Status, Habitat, etc. Resident, patchy, subject to seasonal movements and considerable nomadism. Duars, usually under 500 m: neighbourhood of habitations. Like Indian Myna gregarious, omnivorous and also more or less commensal with man. Habits, behaviour, food and calls very similar to Indian Myna's, the last somewhat softer.

197 JUNGLE MYNA *Acridotheres fuscus* (Wagler)

Size Myna; length 23 cm (9 in.).

Field Characters Very similar to Indian Myna, also with a conspicuous white wing-patch, but distinctly more grey-brown overall. A prominent tuft of erect black feathers at forehead, absence of naked yellow skin round eyes, broader white tips to rectrices, and lemon-yellow iris are other diagnostic clues. Sexes alike.

Status, Habitat, etc. Resident; patchily distributed and subject to local and altitudinal movements. Duars, foothills and up to 2000 m (subspecies *fuscus* and *fumidus*). Less closely associated with human habitation; affects well-wooded country, neighbourhood of forest villages, tea gardens, etc. Habits and behaviour similar to Indian Myna's. Food: chiefly fruits and berries, grain, flower nectar, and insects. Call: high-pitched; like Indian Myna's, including the courtship 'song' accompanied by the comical head-bobbing.

198 HILL MYNA *Gracula religiosa* Linnaeus

Size Myna + ; length 30 cm (12 in.).

Field Characters A stocky, glossy jet black myna with two bright orange-yellow patches of naked skin and fleshy wattles on sides of head and nape. A broad white band on wing-quills very conspicuous in flight. Sexes alike.

Status, Habitat, etc. Resident, subject to slight seasonal local movements. Duars, foothills, locally up to 2000 m: tropical moist-deciduous and semi-evergreen forest (subspecies *intermedia*). Sociable and noisy. Pairs or parties of 5 or 6 keeping to tree tops on forest edges and in cultivation clearings with tall relict trees. Large congregations collect to feed on fruiting *Ficus* trees in company with hornbills, green pigeons and other frugivorous birds. Very lively and noisy towards sunset, settling on bare branches towering above forest, calling and answering one another for long periods before flying off to roost. Food: mainly fruits; also flower nectar, insects and lizards, etc. Call: a wide range of loud whistles, wails, screeches and gurgles, some melodious and very human-like. Much prized as a cage bird for its ability to mimic human speech.

CROWS, JAYS, etc.: Corvidae

199 REDCROWNED JAY *Garrulus glandarius* (Linnaeus) p. 60
Size Pigeon − ; length 35 cm (14 in.).
Field Characters A pinkish brown bird with bright blue wings, closely barred with black, a broad black moustachial band, white rump and vent, and velvety black tail. Sexes alike. In flight fawn-coloured back, glistening white rump contrasting sharply with black tail, diagnostic clues.
Status, Habitat, etc. Resident, rather local, with slight summer-winter altitudinal movements. Normally between 2000 and 3000 m (in winter sometimes down to 1500 m): wet-temperate mixed forest of oak, chestnut, rhododendron and conifers in summer, descending into the semi-tropical zone in winter. A typical crow in habits and behaviour—inquisitive, clannish, and both bold and circumspect as occasion dictates. Usually noisy parties of 4 or 5; larger flocks in winter. Forages on ground as well as trees. Flight seemingly laboured with much flapping of wings. F o o d: fruits and nuts, insects, lizards, young mice and birds, and miscellaneous titbits. C a l l: a harsh loud *shak*; raucous croaks, chuckles and whistling squeals. A low, varied, pleasant 'song', interlarded with good mimicry of other birds' calls.

200 GREEN MAGPIE *Cissa chinensis* (Boddaert) p. 60
Size Myna ± ; length, including long tail, 40 cm (16 in.).
Field Characters A bright leaf-green long-tailed bird with arresting cinnamon-red wings; a black band running backward through eyes to meet on nape. Sexes alike.
Status, Habitat, etc. Resident. Duars and foothills, normally below 1500 m: tropical and subtropical wet evergreen secondary jungle, in overgrown nullahs. Solos, pairs or small noisy parties, usually in association with hunting flocks of laughing thrushes, drongos, etc. in shrubs or on ground—occasionally in tall trees. Behaviour and movements resemble laughing thrushes'. F o o d: mainly animal: large insects, lizards, small birds and eggs; carrion. C a l l: a loud discordant quick-repeated *peep-peep* or *ki-wee*; a raucous mewing note; rich melodious squealing whistles, and mimicry of other birds' calls.

201 YELLOWBILLED BLUE MAGPIE *Cissa flavirostris* (Blyth) p. 60
Size Pigeon ± with tail 40 cm long. Overall length 70 cm (28 in.).
Field Characters A showy long-tailed purplish blue bird with black head, neck, and breast and yellowish underparts. Tail graduated, tipped with black and white, ending in long pliant arching streamers

that trail behind in flight. A small white patch on nape. Bill yellow; legs bright orange. Sexes alike.

Status, Habitat, etc.   Resident, with some summer-winter altitudinal movements. Normally between 2000 and 3300 m in summer; down to 1000 in winter: wet temperate mixed forest of pine, oak, chestnut, rhododendron, etc. (subspecies *flavirostris*). Social and inquisitive. Noisy parties commonly in association with jays and laughing thrushes. Mainly arboreal, but forages also on ground, hopping with tail partly cocked, picking up titbits. Parties fly in follow-my-leader style, tail spread and streamers waving behind. F o o d: practically all-embracing, like jay's; mainly insects and small animals. C a l l: harsh creaking notes and sharp squealing whistles, varied with good mimicry of other birds.

202 **BLACKRUMPED MAGPIE** *Pica pica bottanensis* Delessert   p. 80
Size Myna + with tail 30 cm long. Overall length 50 cm (20 in.).

Field Characters Strongly contrasting black and white plumage and long graduated black tail adequately diagnostic. *Above*, head (all round) and back, including rump, black; scapulars white. Wings and tail black, the former brilliantly glossed with bluish green, the latter with bronze-green and purple. *Below*, breast, thighs, vent and under tail-coverts black; rest pure white. Sexes alike.

Status, Habitat, etc.   Resident; recorded (common) only in Bumthang district of Bhutan between 3400 and 4600 m: open scrub country and cultivation, and groves of trees around upland villages. May occur in northern Arunachal Pradesh adjoining SE. Tibet. Loose parties of 8 to 12. Characteristically crow-like in actions and behaviour: inquisitive, audacious and cunning. Stalks on ground with a strutting gait in search of food, flicking up tail every now and again. Flight rather laboured, flapping and undulating. F o o d: all-embracing—insects, lizards, rodents, weakling birds, etc. C a l l: a subdued rasping *querk* or *kik*. A loud harsh *kekky kekky kekky*, run together as a throaty rattle *chuchuchu* when alarmed.

203 **TREE PIE** *Dendrocitta vagabunda* (Latham)
Size Myna + , with tail 30 cm long. Overall length 50 cm (20 in.).

Field Characters A long-tailed bright rufous arboreal bird with dark sooty head, neck and breast. Broad black tips to graduated greyish tail, and greyish, white and black pattern on wings conspicuous and diagnostic, especially in flight. Sexes alike.

Status, Habitat, etc.   Resident. Duars, foothills, locally up to 2000 m: lightly wooded country in dry- and moist-deciduous biotope—forest plantations, village groves, tea gardens, etc. (subspecies *vagabunda*). A typical crow: social, arboreal and omnivorous; also highly inquisitive, audacious, wary and cunning as occasion dictates. Noisy pairs or parties of 4 or 5, commonly in association with Racket-tailed and other drongos. Flight noisy and dipping—rapid wing flaps alternated with sailing on stiffly outspread wings and tail. Food: fruits, insects, lizards, eggs and weakling birds and miscellaneous small animals; also carrion. Call: a wide range of raucous as well as rich melodious notes: some loud, harsh and rattling like *kā-kā-kā-kā-kāk*, others metallic and flute-like *ko-ki-lā* or *bob-o-link*.

204 **BLACKBROWED TREE PIE** *Dendrocitta frontalis* Horsfield p. 60

Size Myna ± , with tail up to 25 cm long. Overall length 40 cm (16 in.).

Field Characters A pale grey and chestnut tree pie with black face, throat and foreneck, and contrastingly white nape. Heavy bowed black bill. Readily distinguished from the similar Himalayan Tree Pie by entirely black tail (*v.* ashy and black) and absence of large white wing-patch. Sexes alike.

Status, Habitat, etc. Resident, rare. Chiefly foothills and up to 2100 m: dense mixed evergreen forest and bamboo jungle. Habits and behaviour like Himalayan Tree Pie (205), q.v. Small parties. Food: fruit, seeds, insects, lizards, and other small animals. Call: of the characteristic tree pie range of discordant as well as metallic musical notes, cf. 203.

205 **HIMALAYAN TREE PIE** *Dendrocitta formosae* Swinhoe p. 52

Size Myna ± , with tail up to 23 cm long. Overall length 40 cm (16 in.).

Field Characters *Above*, forehead black; sides of head sooty; crown, nape and upper mantle ashy; back and scapulars buffy brown, rump paler. Wings black with a flashing white patch. *Below*, chin and throat dark sooty; rest chiefly rusty and whitish; under tail-coverts chestnut. Sexes alike.

Status, Habitat, etc. Resident, subject to marked summer-winter altitudinal movements. Duars and foothills between 600 and 1500 m, occasionally up to 2300 m: tropical evergreen duar forest or broad-leaved hill forest of oak, rhododendron, etc. (subspecies *himalayensis*). Habits and behaviour typical of tree pies, cf. 203. Foraging parties in fruit-laden trees flutter restlessly in the canopy, flying back and forth with a noisy whirring of wings. When flitting across from one tree top to another actions curiously jerky, with deep saw-edge undulations and closed-wing nosedives. Food: both animal and vegetable. Call: very varied: raucous as well as musical, more or less of same range and pattern as 203, q.v.

206 **NUTCRACKER** *Nucifraga caryocatactes* (Linnaeus) p. 37

Size Pigeon ± ; length 30 cm (12 in.).

Field Characters Crow-like, with stout wedge-shaped bill; chocolate-brown and umber-brown, streaked and spotted with white above and below. Wings black; tail largely, under tail-coverts wholly, white. Sexes alike. In flight the partly spread white tail with a black wedge in middle, and white under tail-coverts, contrast sharply with the dark body and are diagnostic.

Status, Habitat, etc. Resident. Patchily distributed between 2000 and 6000 m: moist-temperate and alpine conifer forest (subspecies *macella*). Pairs or family parties of 4 or 5 in tall conifers on hillsides. Shy and wary, but with characteristic corvine curiosity and inquisitiveness. Attracts notice a long way off by loud grating calls. Flight straight and direct with lazy deliberate wing flaps, recalling jays and blue magpies. At perch rapidly opens and shuts tail, the white rectrices producing a curious flickering effect while bird itself masked by foliage. Very clannish, numbers collecting to join the swearing outcry while investigating an intruder. Food: seeds of conifers, nuts, insects, eggs and nestling birds, etc. Call: harsh, far-carrying, crow-like. A grating *kraak*; sometimes run together as a rattling note.

207 **YELLOWBILLED or ALPINE CHOUGH**
  Pyrrhocorax graculus (Linnaeus) p. 80

Size House Crow − ; length 40 cm (16 in.).

Field Characters A slender glossy jet black crow with yellow bill and bright red legs. Sexes alike.

Status, Habitat, etc. Resident at high altitudes, normally between 2700 and 5000 m, descending in winter to 2400 m: moist- and dry-temperate high mountain biotope with cliffs, alpine meadows, yak pastures and scattered cultivation (subspecies *digitatus*). Very gregarious and sociable, often in large flocks. Occupies a higher altitudinal zone than Redbilled Chough where both species occur. Frequents upland villages, dzongs and traders' encampments to pick up kitchen scraps and titbits. Unlike Redbilled, does not roost or nest in or on monasteries. Flocks disport themselves on thermals, sometimes at great heights in the sky, engaging in remarkable aerobatics with lightning twists and turns and swishing nosedives. Food: insects and grubs, berries, barley, and sometimes carrion. Call: pleasant high-pitched notes *quee-ah* uttered at rest and while frolicking in the air.

208 **REDBILLED CHOUGH**
  Pyrrhocorax pyrrhocorax (Linnaeus) p. 80

Size House Crow ± ; length 45 cm (18 in.).

Field Characters A slender glossy jet black crow with bright red legs and bill, the latter longish and slightly curved. Sexes alike.

Status, Habitat, etc. Resident at high elevations (lower zone than Alpine Chough but largely coincident with it) with appreciable summer-winter altitudinal movements. Between 2400 and 4500 in summer; down to 1600 m in winter (subspecies *himalayanus*): moist- and dry-temperate montane biotope with cliffs, alpine pastures and upland cultivation. Habits and behaviour as in Alpine Chough (207) q.v. Flocks much given to grubbing in meadows and aerial gambolling on thermals in sunny weather. Tame and fearless around dzongs and monasteries, feeding at refuse dumps, roosting under eaves and nesting in holes in walls. Food: insects and grubs, berries, barley grains; apparently not carrion. Call: a shrill, plaintive-sounding *chião*, *chião*; a high-pitched, rather squeaky *khew* and *chee-o-kah*, and others.

209 HOUSE CROW *Corvus splendens* Vieillot
Size Pigeon + ; length 45 cm (18 in.).
Field Characters A medium-sized crow: glossy black with dusky grey nape, neck (all round), upper breast and upper back. Forehead, crown and throat contrasting glossy black. Overall colour pattern very like European Jackdaw. Sexes alike.

Status, Habitat, etc. Resident and local vagrant. Restricted to human habitations, chiefly duars but gradually spreading into the foothills with progressive urbanization. 'Pathfinders' recorded up to 2000 m (subspecies *splendens*). A confirmed commensal of Man, rarely found away from towns and villages. Omnivorous, impudent, bold; intelligent, uncannily wary and untrusting. Has large communal roosts in trees. Food: insects, birds, rodents and other small animals dead or alive; offal and garbage. Highly destructive to eggs and young of small birds. Call: normal note a rather shrill nasal *kaan*, *kaan*; numerous variations of this to suit every mood and situation.

210 JUNGLE CROW *Corvus macrorhynchos* Wagler
Size House Crow + ; length 50 cm (20 in.).
Field Characters A large, uniformly jet black crow with metallic purplish sheen and heavy black bill. Sexes alike.

Status, Habitat, etc. Resident, with slight summer-winter movements at the higher elevations. Duars, foothills and up to 4000 m (subspecies *levaillantii* and *tibetosinensis*). Outskirts of forest villages and outlying homesteads in the lowlands; upland hamlets, labour barracks and shepherds' encampments up to tree-line at the higher elevations. Pairs or small parties; sometimes gatherings up to 50 or so. More a countryside bird, largely independent of Man and his urban concerns. Less

112

Plate 22, artist K. P. Jadav

LAUGHING THRUSHES

K.P.JADAV

0 2 4 6 8 10 cm

Plate 23, artist Winston Creado

YUHINAS, SIVA, SHRIKE-BABBLER, MESIA, BARWING, TIT-BABBLERS

sophisticated than House Crow but equally inquisitive, audacious and cunning. Often associates with choughs at the higher altitudes, and like them extremely fond of frolicking on thermals, performing remarkable aerobatics. Food: all-embracing; both animal and vegetable matter. A habitual marauder of birds' nests. Often associates with vultures at animal carcasses. Call: a deep, hoarse *caw*, more raucous than House Crow's, and also with numerous variations. At higher elevations the Jungle Crow's voice becomes more 'wooden' and raven-like—a guttural *help*, *help*, *help*, etc.

211 RAVEN *Corvus corax* Linnaeus

Size Kite + ; length 70 cm (28 in.).

Field Characters A large heavy jet black high-altitude crow—superficially an enlarged Jungle Crow—with a massive bill. Sexes alike. In overhead flight the wing-quills look slaty, contrasting strongly with jet black under wing-coverts; tail rather wedge-shaped.

Status, Habitat, etc. Resident, normally between 4000 and 5000 m, mostly in typical Tibetan steppe facies; occasionally down to 3000 m in severe winters (subspecies *tibetanus*). Pairs or parties up to a dozen or so around upland habitations and dzongs in dry rocky country above tree-line. Wild, wary, suspicious and inquisitive like rest of family. Feeds in fields and around traders' bivouacs and shepherd encampments; often with Griffon vultures at wayside carcasses. Much given to spectacular aerobatics like choughs. Food: practically every form of vegetable or animal matter, dead or alive. Call: a hoarse, wooden bell-like quick-repeated *prūk*, *prūk*; a high-pitched guttural, almost musical *kreeūk* or *kreeah*.

CUCKOO-SHRIKES, MINIVETS:
Campephagidae

212 PIED FLYCATCHER-SHRIKE *Hemipus picatus* (Sykes)

Size Sparrow; length 14 cm (5½ in.).

Field Characters A small black and white flycatcher-like bird. *Above*, head and back dark smoky brown; rump white. A white collar on hindneck. Tail and wings dark brown-and-white. *Below*, chin, cheeks and throat pure white, running up into hindcollar; rest pinkish white (subspecies *capitalis*). Sexes alike.

Status, Habitat, etc. Resident, with some summer-winter altitudinal movements. Duars, foothills and up to 2000 m: dry- and moist deciduous and evergreen forest, and scrub jungle. Flycatcher-like in behaviour. Makes short sallies after flying insects from tree tops shifting from tree to tree. Gregarious in winter; often among the mixed itinerant hunting parties of small birds. Food: insects. Call: a squeaky, high-pitched *whiriri-whiriri-whiriri* constantly uttered.

213 LARGE WOOD SHRIKE *Tephrodornis gularis* (Raffles)

Size Myna; length 23 cm (9 in.).

Field Characters Male. *Above*, crown and nape ashy grey changing to ashy brown on back, wings and tail; the last tipped with blackish and rufous. A broad black band through eye to whitish ear-coverts. Rump white. *Below*, chin, throat and breast pale ashy fawn, fading to white on belly. Female similar, but crown same colour as back and eye-band brown instead of black.

Status, Habitat, etc. Resident, with some local and altitudinal movements. Duars and foothills up to 1500 m: open moist-deciduous and evergreen secondary forest, and well-wooded country (subspecies *pelvica*). Arboreal. Parties of 4 to 6 often among the roving hunting parties of drongos, minivets and other insectivorous birds. Hunts for insects in tree tops, moving along the branches in bouncing hops; also makes aerial sorties after them. Food: mantids, crickets and similar large insects. Call: some harsh notes like a shrike's; a rather musical *kew-kew* ... quickly repeated four to six times.

214 LARGE CUCKOO-SHRIKE *Coracina novaehollandiae* (Gmelin)

Size Pigeon − ; length 30 cm (12 in.).

Field Characters Male. *Above*, grey, paler on rump; a broad blackish eye-streak. Heavy, slightly hook-tipped bill. *Below*, unbarred whitish. Female (and immature male). *Above*, like adult male, but eye-streak paler. *Below*, chin, throat and breast grey; rest variably barred grey and white (subspecies *nipalensis*).

Status, Habitat, etc. Resident, subject to nomadic and seasonal local movements. Duars, foothills and up to 2400 m in summer, 1800 m in winter: chiefly moist-deciduous openly wooded country and scrub jungle. Pairs or loose parties in tree tops which 'follow-my-leader' from tree to tree. Upon alighting, repeatedly flicks up one wing first and then the other—a characteristically diagnostic action. Food: insects, wild figs, etc. Call: a shrill bisyllabic whistle *tee-treee* or *ti-eee* of parakeet quality (accent on second syllable), chiefly uttered in flight.

215 DARK GREY CUCKOO-SHRIKE
Coracina melaschistos (Hodgson)

Size Myna − ; length 22 cm (8½ in.).

Field Characters Male. Overall dark bluish grey with a prominent blackish eye-stripe. Wing and tail black, the latter broadly white-tipped. **Female** similar but paler grey, often barred on underparts as in immature. A round white patch on underwing and white terminal margin to tail conspicuous in flight, especially when bird alighting. Adult superficially confusable with Plaintive Cuckoo (91), q.v.

Status, Habitat, etc. Resident, subject to summer-winter altitudinal migration, breeding at the higher levels. Duars, foothills and up to 2100 m. Arboreal, usually keeping well up in trees. Singly or pairs, often associated with the mixed itinerant bands of drongos, minivets and suchlike insectivorous birds. Upright stance when perched, and habit of fluttering in front of sprigs to take insects, reminiscent of minivets and wood shrikes. **Food:** chiefly insects and spiders. **Call:** a fairly loud, rather plaintive, 3- or 4-noted whistle *chipee-ee* or *tweet-tweet-tweeor*, uttered from perch or on wing.

216 SCARLET MINIVET *Pericrocotus flammeus* (Forster)   p. 61
Size Bulbul + ; length 23 cm (9 in.).

Field Characters **Male**, showy, slim, glossy black and deep scarlet. *Above*, head, neck and upper back glossy blue-black; lower back and rump deep scarlet; wings and longish graduated tail particoloured black and scarlet. *Below*, throat black; rest bright scarlet. **Female.** *Above*, forehead and short supercilium bright yellow; crown, nape and upper back yellowish grey. Lower back, rump, wings and tail as in male, but with yellow replacing scarlet. *Below*, bright yellow.

Status, Habitat, etc. Resident, with summer-winter altitudinal movements. Duars, foothills, and up to 2700 m (breeding to at least 1800 m): light forest (subspecies *speciosus*). Arboreal. Parties or flocks (up to twenty or more in non-breeding season) often with the mixed hunting parties, flitting restlessly among the canopy foliage. Hovers in front of sprigs or flowers to take insects; also makes aerial flycatcher sallies after winged prey. The flocks 'flow' from tree top to tree top in 'follow-my-leader' style. **Food:** insects and spiders. **Call:** a sweet-toned clear double whistle *twee-twee* repeated every few seconds. Song, a shrill pleasant warble—an elaborate version of the call.

217 SHORTBILLED MINIVET
Pericrocotus brevirostris (Vigors)   p. 64
Size Bulbul − ; length 18 cm (7 in.).

Field Characters **Male**, very similar to Longtailed Minivet; scarcely distinguishable in the field but has a broad scarlet band running through the black wings instead of a large inverted U-shaped patch. **Female**

also like Longtailed, but has more yellow on forehead, extending to crown.

Status, Habitat, etc. Resident, with summer-winter vertical movements; less common than Longtailed Minivet, the ranges of the two largely overlapping. Duars, foothills and up to 2400 m: open forest and well-wooded country. Habits and food similar. Call: aurally indistinguishable.

218 LONGTAILED MINIVET *Pericrocotus ethologus* Bangs & Phillips
Size Bulbul − ; length 18 cm (7 in.).
Field Characters Male. Slim, overall glossy black and deep scarlet-crimson. *Above*, head and mantle glossy black; lower back and rump scarlet-crimson; tail steeply graduated, black-and-crimson; wings black with a large inverted U-shaped crimson patch. *Below*, throat and upper breast glossy black; rest bright scarlet-crimson. Female has grey crown and mantle, yellow throat, and all red parts of male replaced by yellow. Scarcely distinguishable from Shortbilled Minivet (217) in the field.

Status, Habitat, etc. Resident, subject to summer-winter vertical movements. Duars, foothills, and up to 3300 m (subspecies *laetus*). Apparently breeds in a higher zone than Shortbilled Minivet. Habits and behaviour typical, cf. Scarlet Minivet. Gregarious and strictly arboreal. Food: chiefly insects and spiders. Call: a mellow, thin interrogative whistle *weet-twee?* or *swisweet-sweet-sweet* from a perch and in flight.

219 YELLOWTHROATED MINIVET
    Pericrocotus solaris Blyth   p. 64
Size Bulbul − ; length 17 cm (6½ in.).
Field Characters Male. *Above*, crown and back dark slaty; rump and upper tail-coverts deep scarlet; wings and tail black and scarlet. *Below*, chin whitish, throat orange-yellow; rest of lower plumage bright scarlet. Female. *Above*, crown, nape and upper back grey; lower back and rump yellow; wings and tail like male's, but red replaced by yellow. *Below*, chin whitish; rest yellow.

Status, Habitat, etc. Resident, subject to summer-winter vertical movements. Higher foothills; fairly common between 1500 and 3000 m elevation: open tree forest. Essentially a montane species; sociable, arboreal and insectivorous like other minivets. Call: not specifically recorded.

FAIRY BLUEBIRD, IORAS, LEAF BIRDS: Irenidae

220 IORA *Aegithina tiphia* (Linnaeus)
Size Sparrow; length 14 cm (5½ in.).
Field Characters M a l e. A showy black-and-yellow arboreal bird, superficially very tit-like, with two white wing-bars. In non-breeding plumage
similar to female but retaining black tail. F e m a l e yellowish green
throughout; wings and tail greenish brown instead of black; also with
two white wing-bars.
Status, Habitat, etc. Resident at the lower elevations. Duars and foothills
up to 900 m (subspecies *tiphia*): open forest, scrub jungle, groves on
village outskirts. Usually pairs, often in the mixed hunting parties,
hopping from twig to twig, searching the foliage for caterpillars etc.
Actions rather tit-like: frequently clinging to sprigs sideways or upside
down to peer under leaves. F o o d: insects and spiders. C a l l: a variety
of sweet, clear sibilant whistles.

221 GOLDFRONTED CHLOROPSIS
Chloropsis aurifrons (Temminck)   p. 61
Size Bulbul; length 19 cm (7½ in.).
Field Characters An active leaf-green arboreal bird with bright golden-
orange forehead, blue shoulder-patches and slender slightly curved
black bill. Ear-coverts, around eyes and lower throat black; chin and
cheeks (moustachial stripes) dark purplish blue. F e m a l e has the
forehead paler orange-yellow and less extensive.
Status, Habitat, etc. Resident at the lower elevations. Duars, foothills,
sometimes up to 1800 m: moist-deciduous and evergreen biotope—
open, lightly wooded country and secondary scrub. Entirely arboreal;
pairs or small parties. Hunts systematically among the foliage, clinging
to sprigs in all manner of acrobatic positions in the quest. Flight strong
and rapid. F o o d: insects and spiders; berries, and very largely flower
nectar. C a l l: a loud and voluble whistling rattle; a musical drongo-like
swich-chich-chich-wee (accent on first syllable; last much prolonged,
whistle-like). Also excellent mimicry of numerous bird calls given in
quick succession.

222 ORANGEBELLIED CHLOROPSIS
Chloropsis hardwickii Jardine & Selby   p. 65
Size Bulbul; length 19 cm (7½ in.).
Field Characters M a l e. *Above*, leaf-green with a pale greenish blue

shoulder-patch, and deep purple-blue wings and tail. *Below*, chin, throat, sides of head, and breast deep bluish black; bright cobalt-blue moustachial stripes. Belly dull orange. Female has throat, breast, etc. leaf-green like upperparts; orange underparts paler and less extensive; wings largely brown; tail green; moustachial stripes paler blue.

Status, Habitat, etc. Resident, subject to altitudinal movements: the higher foothills and up to 2000 m (generally a higher zone than *aurifrons*)—open forest in moist-deciduous and evergreen biotope. Entirely arboreal. Pairs or small parties in canopy foliage often associated with sibias, spiderhunters, sunbirds and other nectar feeders. Inseparable from the parasitic *Loranthus* clumps infesting lofty forest trees. Very acrobatic in movements while probing into flowers. Food: largely flower nectar; also insects, spiders, and some berries. Call: like Goldfronted, a versatile vocalist but with a longer, sweeter song; also convincing mimicry of other birds' calls.

223 **FAIRY BLUEBIRD** *Irena puella* (Latham)   p. 61
Size Myna + ; length 25 cm (10 in.).
Field Characters Male. *Above*, brilliant ultramarine blue with a silky sheen. *Below*, velvety jet black; under tail-coverts same as back. Female dull blue-green with blackish lores.
Status, Habitat, etc. Resident, subject to seasonal wandering. Duars, foothills and up to 1300 m: tropical and sub-tropical moist-deciduous and evergreen forest—heavy jungle, tea gardens, etc. Pairs or small parties of 6 to 8 usually in tree tops, hopping from branch to branch, flying from tree to tree uttering the distinctive calls. Food: chiefly fruits, berries and flower nectar. Call: a liquid, percussive, whistling double note *weet, weet* or *be-quick* or *peepit* etc., uttered every few seconds.

BULBULS: Pycnonotidae

224 **BLACKCRESTED YELLOW BULBUL**
   Pycnonotus melanicterus (Gmelin)   p. 61
Size Bulbul; length 18 cm (7 in.).
Field Characters *Above*, entire head, throat and erect pointed crest glossy black; rest of upperparts olive-yellow; tail largely brown. Pale yellow eyes conspicuous at short range. *Below*, yellow, darker on breast (subspecies *flaviventris*). Sexes alike.
Status, Habitat, etc. Resident. Duars and foothills normally up to 1500 m: dense secondary jungle and scrub around cultivation and

orchards etc. Arboreal and mainly frugivorous. Usually solos or pairs; larger gatherings on fruiting trees. Often hawks flying ants by aerial sorties from tree tops. Food: chiefly fruits and berries. Call: low churring contact notes. Song, a lively *weet-tre-trippy-weet*, the last 3 syllables repeated twice or more.

225 REDWHISKERED BULBUL *Pycnonotus jocosus* (Linnaeus) p. 61

Size Bulbul; length 20 cm (8 in.).

Field Characters *Above*, dark hair-brown with a pointed black upstanding crest and crimson tuft behind eye; tail brown, with broad white tips to the rectrices. *Below*, white; a brown interrupted collar on upper breast and crimson under tail-coverts (subspecies *monticola*). Sexes alike.

Status, Habitat, etc. Resident. Duars and foothills up to 1100 m: scrub jungle and clearings near villages. Pairs or small parties, occasionally loose flocks of up to a hundred or more where food abundant. Feeds in fruit-laden trees and bushes, but will readily descend and hop about on ground to pick ants. Winged insects captured by aerial sorties from bushtops. Flight jerky but well-sustained. Food: fruits, berries, flower nectar; also insects and spiders. Call: a variety of cheerful musical notes, one of the commonest being a lively *pettigrew* or *kick-pettigrew*. Song, a rollicking phrase syllabified as *The rice must be finished off*.

226 WHITECHEEKED BULBUL *Pycnonotus leucogenys* (Gray)

Size Redvented Bulbul ± ; length 20 cm (8 in.).

Field Characters A sprightly, earth-brown bulbul with long forwardly curving pointed crest, short thin superciliary stripe, glistening white cheek-patches, black throat, yellow under tail-coverts, and white-tipped tail (subspecies *leucogenys*). Sexes alike.

Status, Habitat, etc. Resident, subject to local movements. Duars and foothills, normally between 300 and 1800 m; restricted to the drier valleys in open scrub and bushes, often near habitations. Habits and behaviour very similar to Redwhiskered Bulbul (225) q.v. Food: fruits and berries, flower nectar, insects. Call: similar in pattern to Redwhiskered and Redvented. Song, of 3 or 4 melodious cheerful rollicking phrases like *Tea for two* or *Take me with you*, in endless combinations.

227 REDVENTED BULBUL *Pycnonotus cafer* (Linnaeus)   p. 61

Size Myna − ; length 20 cm (8 in.).

Field Characters A perky earth-brown bulbul with slightly tufted black head, black throat, and scale-like markings on back and breast. White rump, scarlet vent, and white-tipped blackish tail particularly conspicuous in flight (subspecies *bengalensis*). Sexes alike.

Status, Habitat, etc. Resident. Duars, foothills and up to 1800 m: secondary growth, scrub on forest edges, gardens, and around cultivation. Pairs or small parties. Larger gatherings where food seasonally superabundant, e.g. a *Ficus* grove in fruit. Usually at lower elevations than *P. jocosus*, but in some localities position reversed or the two co-existing side by side. Food: fruits and berries, flower nectar, insects and spiders. Call: of similar pattern to Whitecheeked and Redwhiskered bulbuls'. Cheery musical notes *Be quick-quick* and variations.

228 STRIATED GREEN BULBUL *Pycnonotus striatus* (Blyth) p. 65

Size Bulbul ± ; length 20 cm (8 in.).

Field Characters A crested olive-green bulbul with yellow-streaked underparts. *Above*, olive-green finely streaked with white; an upstanding brownish olive crest and yellow rim round eye. Wings and tail largely olive and brown, the latter yellowish on the underside with pale yellow tips to outer feathers. *Below*, throat yellow with dark stipples; breast grey, becoming olive on belly, heavily streaked with sooty and yellow; under tail-coverts yellow (subspecies *striatus* and *arctus*). Sexes alike.

Status, Habitat, etc. Resident between 1200 and 2400 m, descending into lower foothills in winter: medium-sized tree-and-bush jungle in evergreen biotope. Feeding flocks of 6 to 15 in fruit-laden trees and shrubs, 'flowing' from tree to tree in loose follow-my-leader style. Food: mainly berries; also insects, sometimes hawked in air. Call: typical of Redvented and Redwhiskered, but readily distinguishable. Song, a musical *chik-koo* followed by a clear pleasant warbling *weeweewee-weewee*. Also some chattering.

229 WHITETHROATED BULBUL *Criniger flaveolus* (Gould) p. 65

Size Myna ± ; length 23 cm (9 in.).

Field Characters A large crested bulbul. *Above*, olive-green with rufous-brown wings and tail; sides of head grey. *Below*, chin and throat white, rest lemon-yellow. Sexes alike.

Status, Habitat, etc. Resident, subject to seasonal vertical movements. Normally at low elevations, between 600 and 1200 m, descending into lower foothills, duars and terai in winter: dense shrubby undergrowth and secondary jungle in evergreen biotope. Gregarious; rather reminiscent of a laughing thrush. Noisy, chattering feeding parties of 6 to 15 in fruiting shrubs, flying in loose follow-my-leader style from one to the next. Perches with tail well fanned out. Flight strong and direct. Food: berries, wild figs and insects. Call: a loud, harsh, nasal, frog-like croaking *chake chake* (or *kēnk kēnk*), and some sweet and clear notes.

230 RUFOUSBELLIED BULBUL
p. 65
Hypsipetes mcclellandi Horsfield

Size Myna ± ; length 23 cm (9 in.).

Field Characters *Above*, crown dark brown, slightly crested, finely streaked with whitish; rest of upperparts olive-green. *Below*, throat dark grey with heavy white streaks; breast cinnamon, finely streaked with white; belly paler, under tail-coverts yellowish. Tousled crest erected and lanceolate feathers of throat ('beard') stand out when bird is calling (subspecies *mcclellandi*). Sexes alike.

Status, Habitat, etc. Resident, subject to vertical movements, between 900 and 2700 m elevation. Descends in winter to the lower foothills and duars: thick secondary jungle and overgrown cultivation clearings; also oak, rhododendron and open pine forest. Less sociable than other bulbuls. Pairs or small parties frequently with the mixed hunting flocks. Food: mainly drupes and berries. Call: a pleasant clear, sharp *tsyi-tsyi* repeated several times; a full sweet song of three bars constantly uttered in breeding season.

231 BROWNEARED BULBUL *Hypsipetes flavalus* (Blyth)
p. 61

Size Bulbul; length 20 cm (8 in.).

Field Characters *Above*, dark grey with short tuft-like crest. Cheeks black; ear-coverts light brown with a silky sheen. Wings brown, with a large olive-yellow patch. Tail brown, edged with olive-yellow. *Below*, pale grey, white on throat and belly. Sexes alike.

Status, Habitat, etc. Resident. Foothills from 700 up to 1600 m, descending to the duars in winter. Deep secondary forest and tea plantations etc.; fairly open country, edges of forest and cultivation. Forest-loving, arboreal and sociable. Noisy flocks in bushes as well as high trees. Food: berries, flower nectar, insects—the last sometimes captured by aerial sorties. Call: a flute-like note and pleasant jerky song. A soft musical *Daddy, leave-it*; sometimes *Daddy* repeated thrice or 4 times followed by the concluding *leave-it*. Many notes similar to Black Bulbul's (232) q.v.

232 BLACK BULBUL *Hypsipetes madagascariensis* (P. L. S. Müller)
p. 61

Size Bulbul; length 23 cm (9 in.).

Field Characters A slate-grey bulbul with black tuft-like crest, slightly forked tail and bright red bill, legs and feet (subspecies *psaroides* and *nigriscens*). Sexes alike.

Status, Habitat, etc. Resident, subject to summer-winter vertical movements, between 1000 and 3000 m elevation. Descends lower and to the duars in winter: tall forest of oak, pine, rhododendron, etc.; also wooded hill-station gardens; lowland forest in winter. Arboreal, restless, sociable and very noisy, the flocks constantly flying from tree top to tree top in loose disorderly rabbles. Flight strong and swift. Food: mainly fruits, berries and flower nectar; also insects, sometimes hawked by aerial sorties from tree tops. Call: a large variety of loud squeaky notes while chasing one another among the trees; a long-drawn *weenh* and another *whew whé* reminiscent of a creaky gate hinge.

BABBLERS, FLYCATCHERS, WARBLERS, THRUSHES: Muscicapidae

BABBLERS: Timaliinae

233 SPOTTED BABBLER *Pellorneum ruficeps* Swainson   p. 81
Size Bulbul − ; length 15 cm (6 in.).
Field Characters A small terrestrial babbler. *Above*, rufous-brown with a darker chestnut cap, a pale superciliary stripe to behind eye, and brown ear-coverts. *Below*, throat white, rest buff boldly streaked and spotted with blackish (subspecies *mandellii*). Sexes alike.
Status, Habitat, etc. Resident. Duars, foothills and up to 1800 m: heavy brush, secondary growth, bamboo jungle and wooded ravines. Pairs or small parties rummage quietly among dead leaves and mulch on ground. Progresses by thrush-like hops or running like quail. Food: insects. Call: a 3- or 4-noted clear, plaintive ascending whistle, syllabified as *He'll beat you* or *He-will-beat-you* or merely *He'll beat* (or *Pret-ty-sweet*), repeated monotonously from a concealed perch for many minutes. Song, a loud and remarkably beautiful phrase of several rich whistling notes up and down the scale, sustained continuously for a minute or more and soon repeated.

234 MARSH SPOTTED BABBLER *Pellorneum palustre* Gould
Size Sparrow; length 15 cm (6 in.).
Field Characters Superficially similar to Spotted Babbler (233) q.v. *Above*, including tail, uniform (unstreaked) olive-brown. *Below*, chin white; throat and upper breast white streaked with brown; sides of neck, a band across breast, and flanks ochraceous, somewhat streaked with brown; centre of belly white; under tail-coverts ochraceous. Sexes alike.
Status, Habitat, etc. Resident and locally common in Arunachal Pradesh: plains level to 800 m, in elephant grass bordering swamps, and bushes and low jungle on marshy ground. A great skulker, difficult to observe.

When flushed, flits in a lopsided manner and soon dives into the grass again. Food: insects. Call: a loud double *chi-chew*. Song unrecorded.

235 BROWN BABBLER *Pellorneum albiventre* (Godwin-Austen)

Size Sparrow ± ; length 15 cm (6 in.).

Field Characters A small brown short-tailed babbler with whitish faintly spotted throat and whitish belly. Breast not striped but with a rusty wash; also on flanks (subspecies *ignotum*). Sexes alike. Easily mistaken for Spotted Babbler when flitting for cover, but its stubby wren-like tail diagnostic.

Abbott's Babbler (237) is similar but larger, with an unspotted greyish white throat.

Status, Habitat, etc. Resident, subject to short summer-winter vertical movements. Locally common in Bhutan and Arunachal foothills between 500 and 1500 m: scrub and bamboo jungle. A great skulker; shy and retiring. Pairs near the ground in thick tangled growth. Food: insects. Call: a sharp *chick*, some soft chuckling notes, a low clear whistle; a low rippling alarm-note.

236 TICKELL'S BABBLER *Trichastoma tickelli* (Blyth)

Size Sparrow; length 15 cm (6 in.).

Field Characters A long-legged, short-tailed terrestrial babbler. *Above*, olive-brown. *Below*, including throat unstreaked buffish white (subspecies *assamensis*). Sexes alike. Abbott's Babbler (237) is rather similar but larger and with a greyish white throat.

Status, Habitat, etc. Resident, subject to vertical movements. Locally common in Arunachal Pradesh from the Dafla Hills to the Mishmis, plains level to 2100 m: heavy scrub and bamboo jungle. Skulks in dense undergrowth. Feeds on ground. Hops to nearest cover on disturbance, and flies only with reluctance. Food: insects. Call: a loud rapidly uttered *pit-you* ... *pit-you*; a soft rippling *chir-chir* alarm-note.

237 ABBOTT'S BABBLER *Trichastoma abbotti* (Blyth) p. 61

Size Sparrow ± ; length 15 cm (6 in.).

Field Characters A brown short-tailed terrestrial babbler. *Above*, olive-brown. *Below*, throat greyish white; breast and flanks washed with olive; centre of belly whitish, under tail-coverts bright ochraceous. Sexes alike.

Status, Habitat, etc. Resident. Duars, foothills up to 1000 m: tangled thickets in ravines, and broken ground in deep wet jungle. Solitary or pairs in brushwood near ground. Habits and behaviour like Spotted

Babbler, cf. 233. Food: insects. Call: loud, distinctive, of 3 notes dropping in pitch on the middle note; sometimes 4 notes, the first low, the other 3 higher and on the same pitch.

238 SLATYHEADED SCIMITAR BABBLER
Pomatorhinus schisticeps Hodgson

Size Bulbul + ; length 22 cm (8½ in.).

Field Characters *Above*, olive-brown, crown dark slaty; a conspicuous white supercilium to well behind eye; a blackish band through eye to ear-coverts. Longish curved (scimitar-shaped) black-and-yellow bill. *Below*, throat, breast and belly white bordered with chestnut; posterior flanks and under tail-coverts olive-brown (subspecies *schisticeps* and *salimalii*). Sexes alike.

Status, Habitat, etc. Resident. Duars, and foothills up to 1300 m: heavy secondary growth, scrub jungle and mixed bamboo forest. Pairs or small parties. Shy and skulking. Rummages for food among the mulch under shrubbery, moving in long hops. Also hunts on moss-covered branches of trees. Food: insects, seeds, berries. Call: a musical fluty or bubbling *oop-pū-pū-pū* or *oop-pū-pū-pū-pū* uttered by male and almost invariably answered by female with a subdued *kroo-kroo*, the two sounding as a single call.

239 RUFOUSNECKED SCIMITAR BABBLER
Pomatorhinus ruficollis Hodgson   p. 80

Size Bulbul ± ; length 19 cm (7½ in.).

Field Characters *Above*, olive-brown. A conspicuous white supercilium from bill to nape; a black band from lores to ear-coverts; a rufous-chestnut patch on each side of neck. *Below*, only throat white; rest rusty brown, streaked on breast with white. Bill shorter and less curved than in 238. Sexes alike (subspecies *godwini*).

Status, Habitat, etc. Resident in a higher altitudinal zone than other scimitar babblers, 1500 to 3000 m, some descending to 700 m in winter: thickly forested hillsides, rhododendron shrubbery and bracken in summer; forest edges of tea plantations etc. at lower elevations in winter. Habits and behaviour as of other scimitar babblers, cf. 238. Food: mainly insects. Call: a soft musical double- or triple-noted *Off'n on* or *Off-an-on*, reminiscent of Redvented Bulbul's, invariably followed by female's prompt squeak in response; many variations.

240 RUSTYCHEEKED SCIMITAR BABBLER
Pomatorhinus erythrogenys Vigors   p. 61

Size Myna ± ; length 25 cm (10 in.).

Field Characters *Above*, olive-brown; sides of head and neck ferruginous or rusty chestnut; no supercilium; long curved bill and greyish white eyes. *Below*, chin, throat and upper breast dusky grey; lower breast, flanks, thighs and under tail-coverts ferruginous (subspecies *haringtoni* and *mcclellandi*). Sexes alike.

Status, Habitat, etc. Resident, locally common. Duars, foothills and up to 2500 m: dry or moist forest, scrub jungle, secondary growth in ravines, clearings, abandoned cultivation, etc. Pairs or small parties. Feeds on ground flicking aside the leafy litter and digging into damp soil. Sometimes ascends trees with agile thrush-like hops from branch to branch. **Food**: insects, grubs, seeds, berries. **Call**: male's loud, melodious whistle *cue-pee* (accent on *cue*) promptly answered *kip* by accompanying female. Thus actually a duet *cue-pee ... kip* repeated several times, though sounding like a solo.

241 LARGE SCIMITAR BABBLER *Pomatorhinus hypoleucos* (Blyth)

Size Myna $\pm$; length 30 cm (12 in.).

Field Characters A large strong-footed babbler with a long, stout, curved bill. *Above*, olive-brown; a rust-coloured line from eye to nape. Wings and tail brown. *Below*, throat, breast and belly white; flanks slaty, streaked with white on sides of breast. Sexes alike.

Status, Habitat, etc. Resident, locally common in the Dafla and Miri Hills (Arunachal Pradesh). Submontane tract and foothills up to 1200 m: dense bamboo or cane jungle, reeds and elephant grass; sometimes heavy undergrowth in forest. Pairs or small parties. More terrestrial than most scimitar babblers, moving in ungainly hops and very reluctant to fly. A great skulker, oftener heard than seen. **Food**: insects, snails. **Call**: a short chuckle while feeding undisturbed; a loud and full *hoot-hoot-hoot* when alarmed.

242 CORALBILLED SCIMITAR BABBLER

Pomatorhinus ferruginosus Blyth   p. 80

Size Bulbul + ; length 22 cm (8½ in.).

Field Characters *Above*, crown black; a conspicuous white supercilium; a broad black band from lores to ear-coverts. Rest of upperparts, and tail, olive-brown. *Below*, a malar stripe and chin white; throat and breast ferruginous; rest olive-brown (subspecies *ferruginosus*). Sexes alike. Coral-red bill and dark underparts good diagnostic pointers.

Status, Habitat, etc. Resident: normally between 1200 and 3800 m even in winter; also duars and lower foothills—dense shrubbery and ringal bamboo jungle. Pairs or small parties. Elusive and easily overlooked. Rummages among leafy litter and slinks from cover to cover moving in rat-like hops. Flies only when pressed, with alternate flapping and sailing. Food: insects and grubs. Call: musical fluty notes very similar to *oop-pū-pū-pū* of Slatyheaded (238); low grating *churrr* when disappearing on alarm.

243 LONGBILLED SCIMITAR BABBLER
Pomatorhinus ochraceiceps Walden

Size Bulbul + ; length 22 cm (8½ in.).

Field Characters A very slim, long-tailed scimitar babbler with long, deeply curved orange bill. *Above*, back olive, crown and nape more tawny; a conspicuous white supercilium, black lores and dark ear-coverts. Wings and tail brown. *Below*, chin and throat white; breast and belly creamy buff (subspecies *stenorhynchus*). Sexes alike. Very similar to Coralbilled (242) which has dark underparts and shorter, less curved, coral-red bill.

Status, Habitat, etc. Resident; common in the Mishmi Hills (Arunachal Pradesh) above 1500 m, descending to the foothills in winter: dense forest and bamboo jungle. Pairs or parties of a half-dozen or so. Feeds on ground under scrub. Food: mainly insects; also flower nectar. Call: a soft full *hoot-hoot* and a pleasant whistling chuckle; different from other scimitar babblers', but of the characteristic pattern.

244 SLENDERBILLED SCIMITAR BABBLER
Xiphirhynchus superciliaris Blyth p. 80

Size Bulbul; length 20 cm (8 in.) including bill.

Field Characters The disproportionately long and slender curved black bill render identification easy. *Above*, head slaty with a prominent white supercilium; back rufous-brown; wings and tail dark brown. *Below*, throat ashy, lightly streaked with white; rest cinnamon (subspecies *superciliaris* and *intextus*). Sexes alike.

Status, Habitat, etc. Resident at fairly high altitudes, subject to vertical movements; locally common. Breeding between 2100 and 3400 m, descending in winter to between 2000 and 600 m: thick ringal bamboo growth, bramble thickets on steep grassy hillsides, etc. Shy and restless. Pairs or small noisy parties going about in follow-my-leader style. Forages on ground in undergrowth, moving in long rat-like hops. Food: insects, berries, flower nectar. Call: a 3-noted rather interrogative rippling whistle; a single mellow high-pitched *hoot*. A harsh swearing or chittering alarm-note.

245 LONGBILLED WREN-BABBLER *Rimator malacoptilus* Blyth
p. 81

Size Sparrow ± ; length 12 cm (5 in.).

Field Characters A small brown stub-tailed babbler with long slightly

Plate 24, artist J. P. Irani

FLYCATCHERS, SIBIAS, NILTAVAS

1 R   **PALE BLUE FLYCATCHER,** *Muscicapa unicolor*   page 167
    Sparrow ±. ♂ Throat pale blue. Belly whitish. Up
    to 1800 m.

2 R   **BLACKCAPPED SIBIA,** *Heterophasia capistrata*   158
    Bulbul +. Broad black-and-grey terminal band to
    graduated tail. Up to 2500 m.

3 R   **LONGTAILED SIBIA,** *Heterophasia picaoides*   158
    Bulbul. White wing-patch. Graduated whitish-tipped
    tail. Up to 2000 m.

4 R   **VERDITER FLYCATCHER,** *Muscicapa thalassina*   169
    Sparrow ±. Up to 2500 m.

5 R   **RUFOUSBREASTED BLUE FLYCATCHER,**
    Muscicapa hyperythra   163
    Sparrow −. ♀ Olive-brown. Conspicuous whitish
    supercilium (both sexes). Up to 3000 m.

6 M   **REDBREASTED FLYCATCHER,** *Muscicapa parva*   162
    Sparrow −. White patch on each side of tail (both
    sexes). Up to 1500 m.

7 R   **SMALL NILTAVA,** *Muscicapa macgrigoriae*   166
    Sparrow −. Blue patch on each side of neck (both
    sexes). Up to 2000 m.

8 R   **BROOKS'S FLYCATCHER,** *Muscicapa poliogenys*   167
    Sparrow −. Up to 1500 m.

9 R   **RUFOUSBELLIED NILTAVA,** *Muscicapa sundara*   166
    Sparrow. Blue patch on each side of neck (both
    sexes). ♀ White patch on lower throat. Up to 3200 m.

10 R   **BLUETHROATED FLYCATCHER,** *Muscicapa*   168
    rubeculoides
    Sparrow −. Up to 2000 m.

11 R   **SOOTY FLYCATCHER,** *Muscicapa sibirica*   159
    Sparrow −. 1500–3500 m.

12 R   **LITTLE PIED FLYCATCHER,** *Muscicapa*   164
    westermanni
    Sparrow −. Diminutive size. Up to 2500 m.

R. Scholz

Inches
0   2   4

Plate 25, artist Robert Scholz

FLYCATCHERS, NILTAVA, BUSH CHAT, REDSTARTS

1 R ORANGEGORGETED FLYCATCHER, *Muscicapa strophiata*
page 162
Sparrow —. Tail (both sexes) black; white on each side of base. White forehead diagnostic from Redbreasted (323). Lower foothills to 3000 m.

2 R SAPPHIREHEADED FLYCATCHER, *Muscicapa sapphira*
165
Sparrow —. ♀ Rump rusty. Up to 1700 m.

3 R SLATY BLUE FLYCATCHER, *Muscicapa leucomelanura*
165
Sparrow —. Diminutive. Tail: ♂ black, sides of base white; ♀ rufous. Duars to 4000 m.

4 R PIGMY BLUE FLYCATCHER, *Muscicapella hodgsoni*
169
Sparrow —. Diminutive. ♀ Brown above: rufous rump. Pale yellow below. Foothills to 3000 m.

5 R LARGE NILTAVA, *Muscicapa grandis*
165
Bulbul. Blue patch on each side of neck (both sexes). Duars to 2700 m.

6 R YELLOWBELLIED FANTAIL FLYCATCHER, *Rhipidura hypoxantha*
170
Sparrow —. Diminutive. Tail white-shafted, whitetipped; fanned, partially cocked (illustration wrong). Duars to 3600 m.

7 RM COLLARED BUSH CHAT, *Saxicola torquata*
204
Sparrow —. ♀ Dark-streaked rufous-brown above: rump pale rufous. A white wing-patch. Duars to 1300 m.

8 R WHITETHROATED REDSTART, *Phoenicurus schisticeps*
198
Sparrow ±. ♀ Brown; rump chestnut. Ashy wingpatch. White throat-patch as ♂. 1400–4500 m.

9 R GÜLDENSTÄDT'S REDSTART, *Phoenicurus erythrogaster*
199
Sparrow +. ♂ Similar to Whitecapped Redstart (427) but white wing-patch distinctive. 1500–5200 m.

curved bill. *Above*, dark rufescent brown with buff shaft-streaks. A narrow black moustachial stripe. *Below*, chin fulvous; throat, breast and belly pale rufescent brown with whitish shaft-streaks; flanks plain rufescent brown; under tail-coverts ferruginous. Sexes alike.

Status, Habitat, etc. Scarce resident, probably with some vertical movements, between 900 and 2700 m: forest undergrowth and dense scrub in steep broken country. Pairs. A great skulker, difficult to observe. Chiefly terrestrial, feeding on ground like scimitar babblers. When disturbed flies weakly a few metres and dives into thickets. Food. insects. Call: a sweet chirping whistle only recorded.

246 GRANT'S WREN-BABBLER *Napothera epilepidota* (Temminck)

Size Sparrow − ; length 10 cm (4 in.).

Field Characters A small, short-tailed wren-like babbler. *Above*, scaly-patterned dark brown with pale spots on wing-coverts and tips of secondaries; a long pale supercilium. *Below*, throat white, streaked with dark brown; centre of breast and belly whitish; flanks and vent brown (subspecies *guttaticollis*). Sexes alike.

Status, Habitat, etc. Scarce resident. Foothills up to 1500 m, in small glades in dense forest with bracken, mossy boulders, epiphyte-covered fallen tree-trunks, and similar débris. Pairs rummage quietly among fallen leaves, dodging in and out of obstacles like a wren. Very reluctant to fly. Food: insects. Call: alarm-note, a shrill *chir-r-r*; song, a phrase of pleasant low whistling notes.

247 SCALYBREASTED WREN-BABBLER
Pnoepyga albiventer (Hodgson)

Size Sparrow − ; length 10 cm (4 in.).

Field Characters A plump, tailless, squamated little babbler in two colour phases. White phase. *Above*, scaly-patterned olive-brown. *Below*, white, the feathers black-centred giving a scaly effect; flanks squamated olive-brown. Fulvous phase. *Above*, olive-brown spotted with fulvous. *Below*, as in white phase but white replaced by fulvous (subspecies *albiventer*). Sexes alike in both phases. Distinguished from *P. pusilla* (248) only by slightly larger size.

Status, Habitat, etc. Resident, subject to vertical movements. Breeds between 2700 and 3900 m—a higher altitudinal zone than Brown Wren-Babbler (248); descends in winter as low as 600 m. Affects wet ravines with dense undergrowth of ferns, nettles, etc. near streams. Terrestrial, inquisitive; usually solitary. A great skulker, difficult to observe. Creeps through tangled ground vegetation constantly flicking wings like leaf warbler. Food: insects; seeds. Call: (song ?) a loud, squeaky long-drawn *seek ... sik*, rather ventriloquial in effect, repeated unhurriedly several times. Alarm-note, a shrill piercing whistle and a scolding chittering *tsik, tsik*.

248 BROWN WREN-BABBLER *Pnoepyga pusilla* (Hodgson) p. 96
Size Sparrow − ; length 9 cm (3½ in.).
Field Characters An exact miniature of Scalybreasted Wren-Babbler (247) with the same colour dimorphism (subspecies *pusilla*).
Status, Habitat, etc. Resident in a lower altitudinal zone than *albiventer*, mostly between 1500 and 3000 m: wet evergreen forest with tangled ground vegetation of bracken, nettles, etc. Habits and food as of *albiventer*. Call: largely as of 247. Alarm-note, a sharp explosive scolding *chiruk, chiruk*.

249 TAILED WREN-BABBLER *Spelaeornis caudatus* (Blyth) p. 96
Size Sparrow − ; length 10 cm (4 in.).
Field Characters *Above*, dark brown with a scaly appearance. *Below*, throat, breast and flanks ferruginous, spotted with black on latter two; belly slaty, spotted with white. Sexes alike. Distinguished from other wren-babblers by ferruginous throat extending to breast and flanks.
Status, Habitat, etc. Endemic to eastern Himalayas, from E. Nepal through Bhutan, possibly also Arunachal Pradesh. Apparently breeds above 2400 m; descends to 1800 m in winter: damp undergrowth in dense forest—similar habitat to *Pnoepyga*'s. Habits typically wren-like: terrestrial, solitary, restless, skulking and silent. Food: insects. Call: alarm-note, a low quiet *birrh birrh birrh* uttered for long periods when disturbed.

250 MISHMI WREN-BABBLER *Spelaeornis badeigularis* Ripley p. 61
Size Sparrow − ; length 9 cm (4 in.).
Field Characters *Above*, dark brown with a scaly appearance. *Below*, chin whitish; throat chestnut, finely dark streaked; breast, belly and flanks olive-brown conspicuously spotted with white. Distinguished from *caudatus* (249) in having chestnut restricted to throat.
Status, Habitat, etc. Known only from Dreyi in the Mishmi Hills (Arunachal), 1600 m elevation. Solitary, shy and skulking. Food: insects. Call: not recorded.

251 SPOTTED LONGTAILED WREN-BABBLER
Spelaeornis troglodytoides (Verreaux)
Size Sparrow − ; length 10 cm (4 in.).
Field Characters *Above*, umber brown spotted with black and white; a

conspicuous white mark behind eye. Tail and wings narrowly barred. *Below*, throat, breast and belly white; sides of neck and flanks tawny-olive (subspecies *sherriffi*). Sexes alike.

Status, Habitat, etc.   Only recorded from eastern Bhutan between 3000 and 3300 m elevation, in undergrowth in wet temperate forest. More arboreal than other wren-babblers. Clambers about on mossy tree-trunks and bamboo stems up to a couple of metres. Food: insects. Call: a subdued *cheep*; a low song of 4 or 5 notes.

252 SPOTTED SHORT-TAILED WREN-BABBLER
  Spelaeornis formosus (Walden)   p. 96

Size Sparrow − ; length 10 cm (4 in.).

Field Characters *Above*, head, back and wing-coverts olive-brown speckled with white; rump, wings and tail chestnut-brown barred with black. *Below*, cinnamon, densely spotted with white on throat and breast, speckled with black on belly. Sexes alike.

Status, Habitat, etc.   Rare resident, subject to vertical movements, between 2500 and 1200 m: dank rhododendron forest with thick fern ground cover and mossy rocks and fallen tree-trunks etc. Habits in general like other wren-babblers'; little specifically known. Food: insects. Call: a squeaky *seek ... sick* almost identical with that of Scalybreasted Wren-Babbler (247).

253 WEDGEBILLED WREN *Sphenocichla humei* (Mandelli)

Size Bulbul ± ; length 18 cm (7 in.).

Field Characters A stout, heavy-looking wren with powerful legs and feet, and pointed conical bill. *Above*, very dark brown; crown and upper back mottled with golden brown and with fine white shaft-streaks, more conspicuous on forehead; a prominent pale stripe behind eye continued as spots down each side of neck. Lower back, wings and tail finely barred. *Below*, throat and breast dark brown with fine shaft-streaks; centre of belly pale grey; posterior flanks and lower belly mottled with golden brown (subspecies *humei*). Sexes alike.

Status, Habitat, etc.   Very rare resident, probably breeding at high altitudes, moving lower in winter when recorded at 1200 m. Moves in parties of 10 or so in secondary jungle, hunting in the undergrowth like other wren-babblers, but also clinging to rough bark of large trees and climbing to moderate heights like a tree creeper. A skulker and reluctant to fly. Food: insects. Call: unrecorded.

254 REDFRONTED BABBLER *Stachyris rufifrons* Hume   p. 97

Size Sparrow − ; length 12 cm (5 in.).

Field Characters Very similar to Redheaded Babbler (255) but rufous of crown not sharply demarcated from rest of rufous olive-brown upperparts. Distinguished further by white chin, ochraceous throat with the black streaks very faint, and whitish belly (*v*. all pale yellow). Flanks and lower belly ochraceous. A pale grey supercilium (subspecies *ambigua*). Sexes alike.

Status, Habitat, etc. Resident. From the edge of the plains up to 900 m, in dense undergrowth in ravines, bamboo or grassy scrub jungle— deciduous or semi-evergreen biotope. Small restless feeding parties, often mixed with other babblers, creeping through undergrowth and bamboo clumps. Sometimes takes insects on the wing. Food: chiefly insects. Call: mellow musical iora-like 4-noted whistles *whi-whi-whi-whi*, and conversational chittering.

255 **REDHEADED BABBLER** *Stachyris ruficeps* Blyth

Size Sparrow − ; length 12 cm (5 in.).

Field Characters *Above*, crown rufous-brown sharply demarcated from greyish olive-brown back; lores and orbital area pale yellowish. *Below*, chin and throat pale yellow, finely streaked with black; rest pale yellow, tinged with olivaceous on lower belly and flanks (subspecies *ruficeps*). Sexes alike. For distinguishing from Redfronted Babbler see under 254.

Status, Habitat, etc. Resident in a higher zone (1000 to 2700 m) than Redfronted Babbler, the two overlapping at the lower altitudes. Habits as in 254; actions rather tit-like. Food: chiefly insects. Call: indistinguishable from Redfronted Babbler's, q.v.

256 **GOLDENHEADED BABBLER** *Stachyris chrysaea* Blyth   p. 97

Size Sparrow − ; length 10 cm (4 in.).

Field Characters *Above*, forehead, crown and nape golden yellow, the last two streaked with black; lores and a short moustachial stripe black. Back and ear-coverts yellowish olive. *Below*, bright yellow (subspecies *chrysaea*). Sexes alike.

Status, Habitat, etc. Resident, between 1200 and 2600 m: dense bushes, bamboo and wild raspberry undergrowth in humid secondary and evergreen forest and clearings. Skulking, active and restless. Often in the itinerant mixed feeding parties, sometimes up in the foliage canopy at moderate heights, the flocks flowing from tree to tree. Actions when feeding rather tit-like, occasionally like a flycatcher—tail cocked and wings drooping at sides. Food: mainly insects. Call: a low twittering conversational *chirik, chirik, . . .*, rising to louder and shriller notes when agitated. Song of 7 or 8 notes on the same tone.

257 BLACKTHROATED BABBLER *Stachyris nigriceps* Blyth   p. 97
Size Sparrow − ; length 12 cm (5 in.).
Field Characters *Above*, olive-brown; crown blackish, boldly striped with white; a black supercilium extending to nape. *Below*, chin and throat slaty grey bordered by white malar stripes; rest fulvous, tinged with olivaceous on flanks and lower belly (subspecies *nigriceps* and *coei*). Sexes alike.
Status, Habitat, etc.   Resident between 750 and 1800 m: secondary and bamboo jungle in light or dense forest. Parties of 5 to 20, commonly in the mixed roving flocks, feeding actively and flowing from thicket to thicket. Food: mainly insects. Call: alarm-note, an explosive *chhrrri* repeated several times. Song, a low, sweet, rather mournful whistle.

258 YELLOWBREASTED BABBLER
  Macronous gularis (Horsfield)   p. 61
Size Sparrow − ; length 11 cm (4½ in.).
Field Characters *Above*, greyish olive with tawny-olive cap and wings. A pale yellow supercilium. *Below*, pale yellow with fine dark streaks on chin, throat and breast (subspecies *rubricapilla*). Sexes alike.
Status, Habitat, etc.   Resident. Common in the duars and foothills normally below 600 m: light or dense forest undergrowth and bamboo jungle. Parties up to a dozen or more feeding in the foliage canopy and flowing quickly from tree to tree. Actions and behaviour rather tit-like—similar to other small babblers'. Food: mainly insects. Call: a loud mellow *kew-kew-kew-kew* . . . continued for minutes and resumed after a short break; occasionally varied by a harsh *chichoo* or *chrr-chichoo*.

259 REDCAPPED BABBLER *Timalia pileata* Horsfield   p. 81
Size Sparrow + ; length 17 cm (6½ in.).
Field Characters *Above*, forehead white, continued as a streak over eyes; crown chestnut; lores and stout bill black; ear-coverts white, sides of neck slaty grey. Back olive-brown, lightly streaked on upper back. Tail brown, narrowly barred, graduated. *Below*, throat white finely black-streaked; rest buff and olive-brown (subspecies *bengalensis*). Sexes alike.
Status, Habitat, etc.   Resident. Common in the duars and foothills in low-lying swampy tall grass and scrub jungle areas. Small parties of 6 to 8, foraging in low cover and threading their way through the tangles, rarely showing themselves. Food: insects. Call: alarm-note harsh and rasping, similar to Great Reed Warbler's *korchuk*. Song, a combined flute-like trill and whistle of a half-dozen descending notes. Very distinctive.

260 YELLOW-EYED BABBLER *Chrysomma sinense* (Gmelin)
Size Bulbul − ; length 18 cm (7 in.).
Field Characters *Above*, rufescent brown with cinnamon wings; and long, graduated tail. Lores and short supercilium white; conspicuous orange-yellow eye-rim and black bill. *Below*, glistening white and buff (subspecies *sinense*). Sexes alike.
Status, Habitat, etc. Resident. Duars and foothills, usually below 1000 m: secondary growth, tall grass, scrub and bamboo. Parties of 5 to 15, often associated with wren-warblers. Very elusive, seldom exposing itself above tall grass. Occasionally one will clamber up to the top momentarily to utter its cheeping notes and dive into cover again. Flight jerky and undulating. Food: insects and berries; flower nectar. Call: a loud, clear, somewhat plaintive *cheep-cheep-cheep*. A sweet loud whistling song in breeding season *twee-twee-ta-whit-chu* (accent on *whit*).

261 GREAT PARROTBILL *Conostoma aemodium* Hodgson   p. 60
Size Myna; length 30 cm (12 in.).
Field Characters A large rather clumsy grey-brown bird resembling a laughing thrush, with upright carriage and stout orange-yellow bill. *Above*, forehead whitish; lores and supercilium brown; back olive-brown. *Below*, entirely mouse-grey. Sexes alike.
Status, Habitat, etc. Uncommon resident between 2700 and 3600 m, somewhat lower in winter: ringal bamboo and rhododendron bushes. Pairs or small parties often associated with laughing thrushes and resembling them in habits. Feeds in bushes as well as hopping about on the ground. Rather elusive and difficult to see. Food: bamboo shoots, seeds, berries, insects. Call: a characteristic harsh *krrarchah*, *krarch, krarchah*; a clear musical *wheou wheou*. Also noisy chattering like *Turdoides* babblers.

262 BROWN PARROTBILL *Paradoxornis unicolor* (Hodgson)
Size Bulbul; length 20 cm (8 in.).
Field Characters A dull olive-brown babbler-like bird with dark head, conspicuous black eyebrows and very short thick yellow bill. *Above*, head greyish brown, hoary on the sides; a pale eye-ring and dark eye-stripe. Rest of upperparts olive-brown. *Below*, throat and breast greyish brown, rest olive-brown. Sexes alike.
Status, Habitat, etc. Uncommon resident between 2700 and 3400 m, somewhat lower in winter: keeps almost exclusively to dense *Arundinaria* bamboo and dwarf rhododendrons; noisy skulking parties of a half-dozen or so, often in company with Great Parrotbill (261). Sits very upright on bamboo stems; occasionally clings upside down

to mossy branches in search of food. Flight reluctant, fluttering, babbler-like. Food: bamboo and bracken buds, insects. Call: a faint *churr*, *churr*: also a bleating alarm-note.

263 FULVOUSFRONTED PARROTBILL
Paradoxornis fulvifrons (Hodgson)   p. 208

Size Sparrow − ; length 12 cm (5 in.).

Field Characters Recognized by general fulvous coloration and very short globular bill. *Above*, forehead and crown ochraceous; a broad olive supercilium extending to nape with a smaller ochraceous eye-stripe below. Back olive; a rufous patch along wing; tail brown with rufous base. *Below*, throat and breast fulvous, belly whitish grey (subspecies *fulvifrons* and *chayulensis*). Sexes alike.

Status, Habitat, etc. Uncommon resident between 2700 and 3400 m. Confined almost exclusively to patches of dense *Arundinaria* bamboo. Large parties of up to 30 clamber energetically up and down the bamboo stems scrutinizing them from top to bottom for food. Food: bamboo and birch buds, seeds, insects. Call: a continual twittering, and low mouse-like cheep while flitting among the stems.

264 BLACKFRONTED PARROTBILL
Paradoxornis nipalensis (Hodgson)   p. 208

Size Sparrow − ; length 10 cm (4 in.).

Field Characters A tiny orange-brown bird with conspicuous black supercilium, white moustache, black throat and very deep and short yellowish bill. *Above*, crown and ear-coverts ochraceous; a broad white moustache. Back tawny-olive, brighter on rump; wings blackish with rufous and white edges. *Below*, chin and throat black; sides of neck grey; breast pale grey, belly tawny-white (subspecies *humii*, *poliotis* and *crocotius*). Sexes alike.

Status, Habitat, etc. Uncommon resident between 1200 and 3300 m: moist-deciduous and evergreen biotope—oak, rhododendron and bamboo jungle on steep hillsides and in rocky ravines. Parties of up to 30 or so, often with itinerant mixed flocks of tits and small babblers, hunting energetically and flowing in disorderly rabbles from tree to tree. Habits as in 263; also reminiscent of Redheaded Tit (*Aegithalos*). Food: bamboo buds, insects. Call: a continuous high-pitched twittering and loud purring chatter.

265 LESSER REDHEADED PARROTBILL
Paradoxornis atrosuperciliaris (Godwin-Austen)   p. 208

Size Sparrow; length 15 cm (6 in.).

Field Characters *Above*, crown, nape and sides of neck rufous, ear-coverts ochraceous; a pale area round eye. Back rufous olive-brown; wings rufous and brown; tail brown, graduated. *Below* entirely creamy buff (subspecies *oatesi* [Sikkim, ? Bhutan]; *atrosuperciliaris*, with short black supercilium [Arunachal Pradesh]). Sexes alike. *Oatesi* (without black supercilium) easily confused with *P. ruficeps* (266), but is smaller, with blunter bill and more graduated tail.

Status, Habitat, etc. Uncommon resident between 600 and 1500 m, descending to the foothills in winter. Keeps to reed-bamboo and high grass, and scrub jungle. Gregarious and skulking, working through the grass stems, sometimes clinging upside down like tit, only rarely showing itself at the top for a fleeting moment. Very reluctant to fly unless forced. F o o d: vegetable matter, insects, spiders. C a l l: a distinctive wheezy note like the twang of a guitar; a loud chittering on alarm.

266 GREATER REDHEADED PARROTBILL
Paradoxornis ruficeps Blyth

Size Bulbul – ; length 18 cm (7 in.).

Field Characters Very like 265 but larger, with a less graduated tail and deeper ferruginous sides of head. Sexes alike.

Status, Habitat, etc. Resident; locally distributed from plains level up to 1400 m: bamboo, scrub, and dense thickets of reeds and grasses along river banks. Pairs or small parties often associated with other babblers. Habits as of other parrotbills, qq.v. Actions very tit-like but slower. Flight weak and ill-sustained—a few wing flaps followed by a short glide. F o o d: insects, seeds. C a l l: a distinctive squirrel-like chitter interrupted by a series of slowly pronounced double-note *tee-ur*. Also 'like the plaintive bleat of a small kid in distress'.

267 GREYHEADED PARROTBILL *Paradoxornis gularis* Gray

Size Bulbul – ; length 16 cm (6 in.).

Field Characters *Above*, head dark grey; forehead black extending behind as a supercilium to nape; lores and eye-ring white; bill yellow. Rest of upperparts rufous-brown, wings and tail darker. *Below*, white, with a black bib (subspecies *gularis*). Sexes alike. The conspicuous head pattern renders identification unmistakable.

Status, Habitat, etc. Scarce and local resident, chiefly between 900 and 1500 m, descending to the duars and plains in winter: low trees or undergrowth in lofty forest and bamboo jungle. Small parties of 6 to 8. Less of a skulker than other parrotbills, and not so closely dependent on presence of bamboos. Feeding behaviour rather tit-like. F o o d:

vegetable matter, insects. Call: alarm-note a harsh chattering; contact-call of 4 loud notes on the same tone.

268 GOULD'S or BLACKTHROATED PARROTBILL
Paradoxornis flavirostris Gould

Size Bulbul; length 19 cm (7½ in.).

Field Characters *Above*, crown and nape rufous-chestnut; back olive-brown; ear-coverts black; cheeks white; a very deep parrot-like yellow bill. *Below*, chin black; upper throat barred brown and white; throat deep brown, rest of underparts fulvous (subspecies *flavirostris*). Sexes alike. *P. guttaticollis* of Nagaland and Bangladesh is similar (also with black ear-coverts), but lacks the dark brown throat and has whitish underparts.

Status, Habitat, etc. Scarce resident. Duars and foothills up to 1900 m: mostly valleys with plenty of ekra or elephant grass. Small parties of 8 to a dozen or so. Very shy and skulking, seldom exposing itself above the grasses and only momentarily. Its presence within a 'sea' of grass betrayed by the noise made by the mandibles while nibbling the reeds. Food: vegetable matter, insects. Call: an arresting whistle *phew, phew, phew, phuit* rapidly ascending in scale and volume. Also a bleating or mewing cry.

269 JUNGLE BABBLER *Turdoides striatus* (Dumont)

Size Myna; length 25 cm (10 in.).

Field Characters An untidy looking earthy brown bird with creamy white eyes, sickly yellowish bill and legs, and longish graduated tail seemingly loosely fixed. *Above*, greyish brown. Tail dark grey-brown, noticeably cross-barred. *Below*, fulvous ashy with pale streaks on breast (subspecies *striatus*). Sexes alike.

Status, Habitat, etc. Resident, locally common. Duars and foothills up to 700 m: secondary scrub forest, gardens, orchards, etc. Sociable flocks of 6 to 10 or so noisily rummaging for food among dry leaves and litter on the ground. Food: insects, berries, flower nectar. Call: a harsh conversational *kē-kē-kē* ... and loud discordant squeaking and chattering in chorus when excited.

270 GIANT BABAX *Babax waddelli* Dresser

Size Pigeon; slimmer. Length 30 cm (12 in.).

Field Characters A large babbler. *Above*, ashy grey, broadly striped with blackish brown; black moustachial stripes. *Below*, ashy grey streaked with chestnut; belly and vent ashy (subspecies *waddelli*). Sexes alike.

Status, Habitat, etc. Resident in the high Tibetan facies of the northernmost border areas, between 2800 and 4500 m: arid scrub. Parties of 4 or 5 skulking in bushes or hopping about turning over dead leaves. Food: insects and berries. Call: a rapid series of quavering whistles; a pleasant thrush-like song. Also some harsh grating notes.

271 WHITETHROATED LAUGHING THRUSH
Garrulax albogularis (Gould)   p. 112

Size Myna + ; length 30 cm (12 in.).

Field Characters *Above*, olive-brown with tawny forehead, and black lores and eye-rim. Tail graduated, olive-brown, all feathers except central pair broadly white-tipped, showing as a prominent white band when spread in flight. *Below*, throat white bordered by an olive-brown breast-band; belly and vent ochraceous (subspecies *albogularis*). Sexes alike.

Status, Habitat, etc. Common resident between 1800 to 3300 m in summer, down to 900 in winter: dense moist forest and scrubby hillsides. Gregarious, even in breeding season: flocks of 6 to 12 (up to 50 or more in winter) often in the mixed bird associations. Usually feeds on ground hopping about like Jungle Babbler; also in trees. Food: insects, berries, seeds. Call: very noisy: a continual musical chattering *chip chip chip chip* ... and choruses of sibilant squeals and hisses.

272 NECKLACED LAUGHING THRUSH
Garrulax monileger (Hodgson)   p. 112

Size Myna + ; length 25 cm (10 in.).

Field Characters Confusingly like Blackgorgeted (273) q.v. Differentiated chiefly by smaller size, lack of black cheek-stripe, white throat (*v.* buff), narrower black gorget with white of belly running upward along and below it to ear-coverts, absence of black shoulder-patch, and yellowish brown legs (*v.* slate-grey) (subspecies *monileger* and *badius*). Sexes alike.

Status, Habitat, etc. Resident, from the edge of the plains to 1000 m, locally to 1400 m: thick evergreen and moist-deciduous forest with undergrowth of cane brakes etc. Highly gregarious; flocks of 10–25 often in association with Blackgorgeted and other laughing thrushes. Feeds on the ground among litter of dry leaves etc. Food: insects, snails, lizards, berries and other vegetable matter. Call: noisy choruses of hollow-sounding musical whistles.

273 BLACKGORGETED LAUGHING THRUSH
Garrulax pectoralis (Gould)   p. 112

Size Myna + ; length 30 cm (12 in.).

Field Characters Very similar to Necklaced Laughing Thrush (272) but somewhat larger (subspecies *melanotis*). For differences see under that species. Sexes alike.

Status, Habitat, etc. Common resident, from the edge of the plains to 1200 m elevation, locally to 1700: dense forest, secondary growth and bamboo jungle. Sympatric with its 'double', *G. moniliger*, and often in mixed flocks with it. Very noisy. Community display consists of the birds hopping about on ground, flirting and spreading their wings and bowing vigorously to the accompaniment of loud calls. Food: mostly insects and berries. Call: a constant querulous conversational squeaking *week, week, week* ...; a curious human-like piping interlarded with short high whistles, often in a confused chorus; indistinguishable from calls of Necklaced Laughing Thrush.

274 STRIATED LAUGHING THRUSH *Garrulax striatus* (Vigors)
p. 80

Size Myna + ; length 30 cm (12 in.).

Field Characters A large and predominantly arboreal laughing thrush. *Above*, loose mop-like crest dark brown, streaked with white in front; back umber-brown with fine white streaks. Tail chestnut-brown with minute white tips to outer feathers. *Below*, throat and sides of head densely white-streaked; breast and belly brownish grey with paler streaks (subspecies *sikkimensis* and *cranbrooki*). Sexes alike.

Status, Habitat, etc. Common resident, between 1500 and 2700 m descending to the lower foothills in winter: dense forest, ravines and nullahs with abundant undergrowth. Pairs or small noisy parties of 5 to 8, often in association with other birds. Food: insects, berries, seeds. Call: loud, discordant cackling; shrill kite-like conversational squeals; loud, rich musical whistles *O-willyou-willyou-wit* or *wheeyou-youwitoo* and variants.

275 WHITECRESTED LAUGHING THRUSH
Garrulax leucolophus (Hardwicke)   p. 112.

Size Myna + ; length 30 cm (12 in.).

Field Characters A large olive-brown laughing thrush with white crested head, throat and breast and prominent black eye-mask. *Above*, crown and crest white; a broad black stripe through eyes. Back olive-brown; a rufous nuchal collar. *Below*, throat, sides of neck and breast white, bordered by a rufous band continued from nuchal collar; belly olive-brown (subspecies *leucolophus*). Sexes alike.

Status, Habitat, etc. Common resident. Duars, foothills and up to

2000 m: forest with dense undergrowth, secondary scrub and bamboo jungle in broken country. Gregarious and very noisy; flocks of 6 to 12, sometimes 30 or 40, in trees, moving from one to the next in follow-my-leader style. Feeds largely on ground, digging with bill and progressing by bouncing hops. Food: insects, berries, lizards, etc. Call: cackling community choruses, or 'laughter', with much dancing, posturing and fluttering of half-drooped wings.

276 YELLOWBREASTED LAUGHING THRUSH

Garrulax delesserti (Jerdon)   p. 112

Size Myna; length 23 cm (9 in.).

Field Characters An uncrested laughing thrush chiefly rufous olive-brown, dark grey, and yellow. *Above*, crown and nape slate-grey; back olive-brown tinged with rufous; tail largely rufous. *Below*, throat primrose-yellow; sides of breast dark grey (subspecies *gularis*). Sexes alike.

Status, Habitat, etc. Resident, locally common. Foothills of E. Bhutan and Arunachal between 1000 and 1800 m: dense secondary evergreen undergrowth, and bamboo and scrub jungle. Gregarious: flocks of 6 to 15 individuals, sometimes 30 or more. A great skulker, difficult to observe. Feeds on ground, rummaging among the mulch—occasionally in small trees. Food: mostly insects; also berries and seeds. Call: characteristic of the laughing thrushes: shrill squeaks, chattering and· cackling, often in discordant choruses or 'laughter'.

277 RUFOUSCHINNED LAUGHING THRUSH

Garrulax rufogularis (Gould)   p. 80

Size Myna; length 22 cm (8½ in.).

Field Characters *Above*, back black-spotted umber-brown; forehead, crown and ear-coverts black; tail graduated, chestnut with subterminal white-tipped black band. *Below*, chin and under tail-coverts rufous; throat whitish with black sides mingled with white. Breast pale grey, flanks olive-brown, both black-spotted; belly whitish (subspecies *rufogularis*). Sexes nearly alike.

Status, Habitat, etc. Resident, locally common, between 600 and 1900 m, exceptionally higher: dense undergrowth in oak-rhododendron forest, and secondary scrub. Less gregarious than most laughing thrushes; usually pairs or small family parties. A skulker in shrubbery, feeding on ground. Food: insects, berries, seeds. Call: not very noisy: usual chuckles and low conversational chatter of the genus. Loud squeals when alarmed.

278 WHITESPOTTED LAUGHING THRUSH
Garrulax ocellatus (Vigors) p. 80

Size Pigeon; length 30 cm (12 in.).

Field Characters A large laughing thrush conspicuously black-and-white spotted on chestnut-brown upperparts; crown and ear-coverts black. Tail largely chestnut with black subterminal band and white tip. *Below*, throat black; sides of neck and of lower throat cinnamon-rufous; rest buff, mottled with black on breast and flanks (subspecies *ocellatus*). Sexes alike. The similar Tibetan *G. maximus*, probably found in adjacent Arunachal, is distinguishable by its brown crown, rufous throat and longer tail.

Status, Habitat, etc. Resident at high elevations, 2100 to 3400 m: light forest with undergrowth, thick rhododendron shrubbery, etc. Pairs and small parties of 5 to 8, hopping and feeding on ground. Food: insects, fruits, seeds. Call: shrill, far-carrying notes reminiscent of those of Hawk-Cuckoo (*Cuculus sparverioides*). A beautiful 8-noted piercing whistling song of human quality in breeding season.

279 GREYSIDED LAUGHING THRUSH
Garrulax caerulatus (Hodgson) p. 80

Size Myna ± ; length 25 cm (10 in.).

Field Characters The only laughing thrush in its range with white underparts and grey flanks. *Above*, forehead and orbital area black; ear-coverts whitish; crown rufous-brown, nape olive-brown fulvous on sides, both finely black-barred giving the head a scaly appearance. Mantle and sides of throat olive-brown; wings rufous-brown; tail chestnut. *Below*, white; flanks slaty grey (subspecies *caerulatus*). Sexes alike.

Status, Habitat, etc. Resident, locally distributed between 1500 and 2700 m, descending lower in winter: undergrowth in forest, and ringal bamboo and scrub-covered hillsides. Parties of 3 to 12 or more, skulking in bushes, feeding on the ground. Flight—weak and ill-sustained—and other habits and behaviour typical of the genus. Food: insects, berries, seeds. Call: a constant flow of soft, pleasant conversational notes. Also loud musical liquid whistles, and bursts of discordant cackling.

280 RUFOUSNECKED LAUGHING THRUSH
Garrulax ruficollis (Jardine & Selby) p. 112

Size Myna; length 23 cm (9 in.).

Field Characters A dark laughing thrush with black forehead, ear-coverts, throat and upper breast; crown and nape slaty. A large rufous patch on each side of neck, very prominent when bird calling. Back, rump,

belly, and wings dark olive-brown, the last with pale outer edges.
Vent rufous. Tail black. Sexes alike.

Status, Habitat, etc. Common resident. Duars, foothills and up to
1500 m: bamboo jungle, outskirts of forest, secondary growth, tea
gardens, etc. Pairs or noisy parties of 3 to 20 according to season.
Habits and behaviour typical of the genus. Food: insects, molluscs,
berries, seeds. Call: a large repertoire of squeals and sharp musical
notes, often in chorus. One common call *weeeoo-wihoo-wick* (possibly
a duet) repeated unvaryingly for many minutes with a short pause
between each.

281 STREAKED LAUGHING THRUSH

Garrulax lineatus (Vigors)   p. 112

Size Bulbul $\pm$; length 20 cm (8 in.).

Field Characters A small, uncrested, streaked laughing thrush with a
greyish white terminal band on graduated rounded tail. *Above*, crown
and upper back grey streaked with dark brown; mantle streaked with
white; rump olive-brown; ear-coverts and wings rufous; tail olive-
brown, faintly barred, tipped greyish. *Below*, olive-brown streaked on
throat and breast with rufous (subspecies *setafer* and *imbricatus*).
Sexes alike.

Status, Habitat, etc. Common resident, with some summer-winter
altitudinal movements; breeds between 1200 and 3000 m. Affects bush-
covered hillsides, wooded nullahs, open forest and hill-station gardens,
etc. Pairs or parties of a half-dozen or so. More terrestrial than most
other laughing thrushes. Shuffles or hops along the ground through
tangles of grass, bracken and low undergrowth, flirting wings and jerking
tail. Very reluctant to fly. Food: insects, berries, seeds. Call: incessant
conversational querulous squeaking; loud clear whistles *p'ty weer* or
titty-titty-we are, and variants. Song, a jingling squeaky whistle of
three plaintive descending notes *pee-pi-pi* of timbre reminiscent of
White-eye (*Zosterops*).

282 BLUEWINGED LAUGHING THRUSH

Garrulax squamatus (Gould)   p. 80

Size Myna + ; length 25 cm (10 in.).

Field Characters Dark olive-brown with black scale-like markings over
entire body, especially on back. A black supercilium and white eye very
conspicuous. Wing black with pale blue outer edge and large rufous
shoulder-patch. Tail blackish with rufous terminal band; upper and
under tail-coverts chestnut. Sexes alike.

Status, Habitat, etc. Resident, uncommon: humid dense bushes, ringal

Plate 26, artist J. P. Irani

WARBLERS, FLYCATCHERS, TAILOR BIRD

1 R BLACKTHROATED HILL WARBLER, *Prinia atrogularis*   page 180
Sparrow – . In winter throat black-streaked whitish; a white
supercilium. 1000–2500 m.

2 R PARADISE FLYCATCHER, *Terpsiphone paradisi*   171
Bulbul. Up to 800 m.

3 R GREYHEADED FLYCATCHER, *Culicicapa ceylonensis*   169
Sparrow – . Up to 3000 m.

4 R BLACKNAPED MONARCH FLYCATCHER,
Monarcha azurea   171
Sparrow. ♀ Crown blue; rest of upperparts brown. No black
nape-patch or gorget. Up to 1000 m.

5 R PLAIN WREN-WARBLER, *Prinia subflava*   179
Sparrow – . Supercilium and outer rectrices whitish. Up to
1200 m.

6 R WHITETHROATED FANTAIL FLYCATCHER,
Rhipidura albicollis   170
Bulbul – . Up to 2000 m.

7 R BROWN HILL WARBLER, *Prinia criniger*   179
Sparrow – . Like 1, but upperparts streaked. Underparts buff.
No black throat. 300–2800 m.

8 R RUFOUSCAPPED BUSH WARBLER, *Cettia brunnifrons*   174
Sparrow – . Crown chestnut, back brown. 2200–4000 m
(summer).

9 R ABERRANT BUSH WARBLER, *Cettia flavolivacea*   174
Sparrow – . Crown and back both olive-brown. 700–3600 m.

10 R TAILOR BIRD, *Orthotomus sutorius*   180
Sparrow – . Rusty crown. Pointed graduated tail. Up to
1800 m.

11 R RUFOUS WREN-WARBLER, *Prinia rufescens*   178
Sparrow – . Graduated black-and-white-tipped tail. Up to
2000 m.

12 R ASHY-GREY WREN-WARBLER, *Prinia hodgsonii*   178
Sparrow – . Like 11, but with cloudy grey breast-band
(summer). Up to 1200 m.

13 R YELLOWBROWED GROUND WARBLER, *Tesia cyaniventer*   172
Sparrow – . Diminutive. Stub-tailed. 1800–2500 m (summer).

JPIrani 1972

0   3   6   9 cm

R. Scholz

Inches

Plate 27, artist Robert Scholz

WARBLERS, TAILOR BIRD, SUNBIRDS, FLOWERPECKER

1 R BLACKBROWED FLYCATCHER-WARBLER,
Seicercus burkii   page 187
     Sparrow —. Duars to 3700 m.

2 R CHESTNUTHEADED FLYCATCHER-WARBLER,
Seicercus castaniceps   188
     Sparrow —. 2 yellow wing-bars. White eye-ring.
     Duars to 2400 m.

3 R BLACKFACED FLYCATCHER-WARBLER,
Abroscopus schisticeps   188
     Sparrow —. Broad yellow supercilia meeting on fore-
     head. Broad black band through eye. 1500–2500 m.

4 R PALEFOOTED BUSH WARBLER, *Cettia pallidipes*   173
     Sparrow —. Pale supercilium. Dark stripe through
     eye. Duars to 1500 m.

5 R GOLDENHEADED TAILOR BIRD, *Orthotomus*
cucullatus   180
     Sparrow —. Crown rufous. Short yellow supercilium.
     Dark grey nuchal collar. Breast grey. Rest underparts
     and rump bright yellow. Duars to 1800 m.

6 R FIRETAILED YELLOWBACKED SUNBIRD,
Aethopyga ignicauda   234
     Sparrow —. Back and tail, with elongated central
     rectrices, bright red. ♀ Olive. No yellow band across
     rump. Tail short. 1200–4000 m.

7 R MRS GOULD'S SUNBIRD, *Aethopyga gouldiae*   233
     Sparrow —. Sides of head crimson. Elongated central
     rectrices metallic purple-blue. ♀ Crown grey. Rump
     yellow. Belly yellow; throat grey. Tail short. Duars
     to 3300 m.

8 R BLACKBREASTED SUNBIRD, *Aethopyga saturata*   233
     Sparrow —. Head dark. Elongated central rectrices
     metallic purple. ♀ Olive. Grey crown; yellow rump-
     band; short tail. 450–2000 m.

9 R FIREBREASTED FLOWERPECKER, *Dicaeum*
ignipectus   232
     Sparrow —. ♀ Olive-green; rump yellower. Buff below,
     tinged olive on sides. 750–3000 m.

bamboo and rhododendron scrub, especially along streams. Breeds
between 1000 and 2400 m. Pairs or small parties. An inveterate skulker
with typical laughing thrush habits. Food: insects, berries, seeds.
Call: a thrush-like *chuck*; song rendered as *cur-white-to-go* and *free-
for-you*.

283 PLAINCOLOURED LAUGHING THRUSH
Garrulax subunicolor (Blyth)   p. 112
Size Myna; length 23 cm (9 in.).
Field Characters A scaly-patterned olive-brown laughing thrush like
G. squamatus (282). Distinguished from it by lack of black supercilium,
presence of a large straw-coloured wing-patch, and white-tipped outer
tail-feathers (subspecies *subunicolor*). Sexes alike.
Status, Habitat, etc. Resident, subject to some vertical movements.
Breeds between 1800 and 3600 m. Affects thickets of wild raspberry,
dwarf rhododendron, bamboo and secondary mixed deciduous forest.
Flocks of 10 to 20 birds keeping to tangles of bushes and vines. Habits
typical of the genus. Food: insects, berries; also molluscs, centipedes,
etc. Call: a clear whistle of 4 notes; some squeaky conversational notes,
and a sharp alarm-note.

284 BLACKFACED LAUGHING THRUSH *Garrulax affinis* Blyth
p. 112

Size Myna; length 25 cm (10 in.).
Field Characters Black face and white patches on each side, combined with
yellow-and-slate wings and tail, identifies this species. *Above*, head
mostly black with a white malar patch, white sides to nape, and white
semi eye-ring. Mantle brown, finely scalloped. Wings olive-yellow with
slaty tip and outer edge, and a small black shoulder-spot. Tail olive-
yellow with slaty tip. *Below*, chin black; rest rufous-brown with grey
scale-like scalloping (subspecies *bethelae*). Sexes alike.
Status, Habitat, etc. Common resident, at high altitudes between 2400
and 4200 m, subject to vertical movements: dwarf rhododendron,
scrub and oak and ringal bamboo in mixed oak-conifer-birch forest,
often above timber-line. Pairs or small parties. Noisy and excitable.
Habits as of the genus. Food: insects, berries. Call: a melodious
whistling *to-wee* or *to-wee-you*; the usual conversational chuckles.
Alarm-note, a long rolling *whirr-whirrer*.

285 REDHEADED LAUGHING THRUSH
Garrulax erythrocephalus (Vigors)   p. 80
Size Myna ± ; length 30 cm (12 in.).

Field Characters Olive-brown with rufous-chestnut crown and nape, black chin and scale-like black markings on breast, neck, and upper back. Wings and sides of tail olive-yellow; a chestnut shoulder-patch. Underparts deep ferruginous (subspecies *nigrimentus* and *imprudens*). Sexes alike.

Status, Habitat, etc. Common resident, between 1200 and 3000 m, subject to vertical movements: thick undergrowth in forest, especially tangled bushes in steep-sided ravines. Pairs or parties of 4 to 6, sometimes 30 or more, in company with other laughing thrushes. A great skulker, with habits typical of the genus. Food: insects and berries. Call: constant conversational chuckles; a clear double or treble whistle *tweeyoo* or *tuweeyoo*, and others.

286 CRIMSONWINGED LAUGHING THRUSH
Garrulax phoeniceus (Gould) p. 80
Size Myna ± ; length 23 cm (9 in.).

Field Characters Olive-brown, with bright crimson sides of head, wings and tips of under tail-coverts. Crown streaked with black; a black supercilium. Tail black with reddish tip and outer feathers (subspecies *phoeniceus*). Sexes alike.

Status, Habitat, etc. Resident, locally common between 900 and 1800 m, subject to slight vertical movements (down to the duars in winter): undergrowth in evergreen forest and dense thickets of secondary growth on edge of cultivation. Pairs or small parties of 4 or 5, sometimes associated with other laughing thrushes. A skulker, with habits typical of the genus. Food: insects, berries, seeds. Call: squeaky conversational notes. Song, of 5 or 6 notes, the last 3 or 4 ending on the same tone.

287 SILVEREARED MESIA *Leiothrix argentauris* (Hodgson) p. 113
Size Sparrow; length 15 cm (6 in.).

Field Characters A bright-coloured arboreal babbler with black crown and moustachial stripes and silvery ear-coverts. Male. Forehead yellow; throat and breast bright orange-yellow. Wings edged with yellow, and with a crimson patch. Upper and under tail-coverts crimson. In female upper tail-coverts olive-yellow, under ochraceous (subspecies *argentauris*).

Status, Habitat, etc. Resident, subject to summer-winter vertical movements. Foothills and up to 2100 m: secondary growth, abandoned cultivation clearings, tea gardens, etc.—evergreen biotope. Small parties and flocks of up to 30 or more, often in the mixed itinerant bands of other small babblers. Hunts insects among foliage with acrobatic tit-like actions, the flocks flowing from tree to tree in follow-my-leader fashion. Food: insects, seeds, berries. Call: a continual conversational chirrup while foraging; a long-drawn clear whistling *seesee-siweewee*; a cheerful song of 7 or 8 notes with frequent flirting of wings.

288 REDBILLED LEIOTHRIX or PEKIN ROBIN
Leiothrix lutea (Scopoli)   p. 81

Size Sparrow − ; length 13 cm (5 in.).

Field Characters A sprightly bright-coloured bird, overall greyish olive with bright yellow throat and breast, a pale eye-ring and scarlet bill. Wings black with yellow-and-crimson edges. In female crimson of wings replaced by yellow (subspecies *calipyga*). Distinguished from Silvereared Mesia (287) by olive crown (*v.* black) and absence of silvery white ear-coverts.

Status, Habitat, etc. Resident, locally common. Breeds between 1500 and 2400 m; descends lower in winter. Evergreen biotope: secondary growth, overgrown clearings, tea plantations, etc. Pairs or parties of 4 to 6 sometimes 20 or more, usually associated with roving flocks of other small babblers etc. Habits similar to those of Silvereared Mesia; like it also sometimes feeds on the ground. Food: insects, berries, seeds. Call: alarm-notes, a harsh hissing. Song, a loud cheerful warbling reminiscent of Redwhiskered Bulbul's, more prolonged and musical.

289 FIRETAILED MYZORNIS *Myzornis pyrrhoura* Blyth   p. 97

Size Sparrow−; length 12 cm (5 in.).

Field Characters Male. A brilliant dark green babbler with red-and-green tail. Crown scalloped with black; a black stripe through eye. Wings black with a reddish streak, white tip and white inner edge. Female similar but red tinge of underparts duller, and red in wings and tail less brilliant.

Status, Habitat, etc. Resident, locally distributed, from 1600 to 3600 m: rhododendron, juniper and bamboo thickets, preferably on sunny hillsides. Parties of 3 or 4, or small flocks, often with other small babblers, sunbirds, etc. on flowering shrubs. Hovers at sprigs like sunbird to take an insect; runs up moss-covered tree-trunks like creeper. Food: insects, spiders, berries, flower nectar. Call: normally silent. A high-pitched *tsit-tsit*. Song not recorded.

290 CUTIA *Cutia nipalensis* Hodgson   p. 81

Size Bulbul; length 20 cm (8 in.).

Field Characters A dumpy, short-tailed arboreal babbler with white underparts and bold black rib-like marking on flanks. Male. Crown slaty blue; a broad black band through eye to nape. Back rufous. Female has the eye-band chocolate-brown; back dull rufous with oval black spots (subspecies *nipalensis*).

Status, Habitat, etc.   Resident, rather local, from 1350 to 2500 m: heavy oak and mossy evergreen forest. Small parties of 8 to 10 or so, often associated with mixed itinerant flocks feeding in the foliage canopy. Runs swiftly along branches and hops up mossy tree-trunks; occasionally descends to ground. Food: insects, molluscs, berries, seeds. Call: a loud, *chichip-chip-chip* monotonously repeated for long periods. Usually silent.

291 RUFOUSBELLIED SHRIKE-BABBLER

Pteruthius rufiventer Blyth   p. 81

Size Bulbul; length 17 cm (6½ in.).

Field Characters Male. *Above*, black and chestnut. *Below*, throat and breast ashy; a yellow patch on each side of breast. Rest vinous-brown, darker on flanks. Female. *Above*, grey, black, greenish yellow and chestnut. Tail mostly black, narrowly tipped chestnut. *Below*, as in male.

Status, Habitat, etc.   Resident, uncommon, between 1500 and 2500 m: dense moss-covered oak and evergreen forest, occasionally in secondary scrub. Small parties in company with other babblers and tits feeding near ground as well as in tall trees. Rather lethargic. Food: mostly insects. Call: not satisfactorily known.

292 REDWINGED SHRIKE-BABBLER

Pteruthius flaviscapis (Temminck)   p. 81

Size Myna − ; length 16 cm (6 in.).

Field Characters A stocky short-tailed black-and-white arboreal babbler with conspicuous chestnut in wings. Male. *Above*, head black with a broad white stripe behind eye; back ashy grey. Wings black and chestnut, tipped with white. Tail black. *Below*, ashy white; flanks vinous-brown. Female. *Above*, head grey; back brownish grey. Wings chestnut and black, edged with yellowish green. *Below*, pale buff (subspecies *validirostris*).

Status, Habitat, etc.   Resident, fairly common. Breeds between 1500 and 2500 m; descends lower in winter: heavy broad-leaved forest of oak, rhododendron, etc. Pairs or parties of 6 to 10 usually in the mixed foraging assemblages. Runs swiftly along boughs, hops from branch to branch, clings sideways to trunks like nuthatch, exploring nooks and crannies. Movements sluggish; flight jerky and dipping with hurried wing-beats. Food: insects, berries, seeds. Call: harsh, grating *churr*s while feeding; a loud distinctive *kewkew-kewkew* quickly repeated 3 or 4 times.

293 GREEN SHRIKE-BABBLER *Pteruthius xanthochlorus* Gray p. 113
Size Sparrow − ; length 13 cm (5 in.).
Field Characters M a l e. *Above*, head grey; back olive-green. Closed wings greenish, with blackish shoulder-patch and a faint pale wing-bar. Tail with narrow white tip. *Below*, throat and breast ashy; belly yellow. **Female** similar but duller (subspecies *xanthochlorus*).
Status, Habitat, etc. Resident, uncommon, subject to some vertical movements. Breeds between 1800 and 3000 m, descends lower in winter: forests of oak, spruce, hemlock and deodar. Pairs or parties of 3 or 4 usually amongst the mixed itinerant bands of tits, leaf warblers, etc. Easily mistaken for a *Phylloscopus* but is comparatively sluggish and does not nervously flick its wings. F o o d : insects, berries, seeds. C a l l : a single *whit*; song, a single note rapidly and monotonously repeated.

294 CHESTNUT-THROATED SHRIKE-BABBLER
 Pteruthius melanotis Hodgson   p. 97
Size Sparrow − ; length 11 cm (4½ in.).
Field Characters M a l e. *Above*, olive-green with yellow forehead and grey nuchal collar. A conspicuous white eye-ring and greyish supercilium; a black crescentic line behind yellow ear-coverts. Wings grey with a broad black bar between two narrow white ones. Tail greenish black with white outer rectrices and tips. *Below*, throat and upper breast chestnut, rest yellow. **Female** has head markings less distinct; throat mostly buffish with cinnamon 'moustache'; wing-bars salmon-rufous instead of white (subspecies *melanotis*).
Status, Habitat, etc. Uncommon resident, between 1800 and 2700 m, descending lower in winter: open glades etc. in deep evergreen forest. Entirely arboreal. Pairs or small groups usually among the mixed foraging assemblages of small babblers, flycatchers, warblers, etc. F o o d : chiefly insects. C a l l : contact-note, a pleasant *too-weet*, *too-weet*. Usually silent.

295 WHITEHEADED SHRIKE-BABBLER
 Gampsorhynchus rufulus Blyth   p. 81
Size Bulbul + ; length 23 cm (9 in.).
Field Characters Olive-brown, with entire head, throat, breast and belly white, the last buffish. Wing with a white shoulder-patch and buff inner edge. Tail graduated, tipped with buff (subspecies *rufulus*). Sexes alike.
Status, Habitat, etc. Resident, from edge of plains through duars and foothills up to 1200 m: secondary and bamboo jungle, and undergrowth

in evergreen forest in broken country. Arboreal, gregarious and noisy.
General appearance, flight and behaviour reminiscent of bulbuls.
Food: chiefly insects. Call: a weird grating *kaw-ka-yawk* and others,
constantly uttered.

296 HIMALAYAN BARWING *Actinodura egertoni* Gould   p. 113

Size Bulbul + ; length 23 cm (9 in.).

Field Characters *Above*, forehead rufous; mop-like ashy brown crest.
Back and rump rufous-brown; wings narrowly cross-barred, and with a
black patch within a large rufous patch. Tail graduated, rufous-brown
narrowly cross-barred, tipped white. *Below*, chin rufous; throat and
breast pinkish brown. Rest tawny olive, white on belly (subspecies
egertoni and *lewisi*). Sexes alike.

     A. nipalensis (297) and *A. waldeni* (298) have a darker crown and
black tail.

Status, Habitat, etc. Resident, common. Breeds between 1200 and
2000 m; moves lower in winter: dense secondary evergreen growth.
Pairs or parties 6 to 12. Clambers about in bushes and trees poking
into holes and crannies and amongst epiphytes with tit-like actions,
clinging upside down, sometimes fluttering at sprigs. Food: insects,
berries, seeds. Call: a feeble conversational *cheep*. Song, a three-noted
whistle.

297 HOARY BARWING *Actinodura nipalensis* (Hodgson)   p. 81

Size Bulbul; length 20 cm (8 in.).

Field Characters Differs from Himalayan Barwing (296) by dark brown
crown (*v.* ashy brown), shorter and blackish tail (*v.* rufous-brown),
grey breast (*v.* pinkish brown) (subspecies *vinctura*). Sexes alike.

Status, Habitat, etc. Resident, common, from 2100 to at least 3000 m:
oak, fir and rhododendron forest with dense undergrowth. Habits,
food and voice as in 296, q.v.

298 ARUNACHAL BARWING *Actinodura waldeni* Godwin-Austen

Size Bulbul; length 20 cm (8 in.).

Field Characters Very similar to Hoary Barwing (297) but underparts
grey with rufous-brown streaks on throat and breast *v.* plain grey
(subspecies *daflaensis*). Sexes alike. Distinguished from Himalayan
Barwing (296) by shorter and black tail and lack of white on belly.

Status, Habitat, etc. Common resident from the Dafla to the Mishmi
hills between 2400 and 3300 m, lower in winter: mossy evergreen and
mixed forest. General habits, food and voice as in 296, q.v.

299 REDTAILED MINLA *Minla ignotincta* Hodgson   p. 65
Size Sparrow; length 14 cm (5½ in.).
Field Characters A small bright-coloured babbler. M a l e. *Above*, head
black with a long white supercilium to nape; back chocolate-brown.
Wings black with white tip and crimson outer edge. Tail black with
crimson outer edge and tip, and a white patch at base. *Below*, chin and
throat whitish, rest pale yellow. F e m a l e like male but back olive-
brown; red in wings and tail and yellow of underparts paler (subspecies
ignotincta).
Status, Habitat, etc. Common resident, from 1800 to 3100 m, reaching
the duars and lower foothills in winter: humid dense oak, rhododendron
and mixed evergreen forest. Arboreal and sociable. Usually in large
parties among the mixed roving flocks of other small babblers,
flycatchers, etc. feeding tit-like in high-canopy foliage and flowing from
tree to tree. F o o d: chiefly insects. C a l l: a high-pitched *chik* or *tsi*
quickly repeated 7 or 8 times.

300 BARTHROATED SIVA *Minla strigula* (Hodgson)   p. 81
Size Sparrow; length 14 cm (5½ in.).
Field Characters A small brightly coloured babbler, largely yellow. *Above*,
slightly tufted crown orange-brown; a pale yellow eye-ring and post-
ocular stripe. Back greyish olive. Wing with bright orange outer edge
and a black shoulder-patch; secondaries ashy and black, tipped with
white. Tail black and chestnut, edged and tipped with bright yellow.
Below, chin orange; a black malar stripe; throat whitish, narrowly
black-barred; rest yellow (subspecies *strigula* and *yunnanensis*). Sexes
alike.
Status, Habitat, etc. Common resident. Breeds between 1800 and 3600 m;
winters at lower elevations: oak, rhododendron and mixed forest, and
bamboo jungle. Arboreal; restless parties of 6 to 20, usually among the
mixed itinerant foraging flocks. Habits and food as in 299. C a l l: a
mellow *peera-tzip* or loud *pe-eo*. Song, a prolonged jumble of sweet
whistling notes intermingled with harsh squeaks and *churr*s.

301 BLUEWINGED SIVA *Minla cyanouroptera* (Hodgson)   p. 113
Size Sparrow; length 15 cm (6 in.).
Field Characters *Above*, crown slightly tufted, dark blue striped with
whitish; supercilium and eye-ring white. Back fulvous, paler on rump.
Wings blue with a white spot and white tips. Tail grey-and-blue showing
white outer rectrices when spread. *Below*, pale vinous grey, whitish on
belly (subspecies *cyanouroptera*). Sexes alike.
Status, Habitat, etc. Resident, locally common, between 1200 and

2200 m, descending to the lower foothills and even duars in winter: evergreen forest, secondary growth and mixed bamboo jungle. Arboreal and gregarious: parties of 5 to 15 usually among the mixed itinerant flocks of other small babblers, flycatchers, etc. General habits like 299. Food: mostly insects. Call: a chick-like *cheep* or *cree-cree*. Song, a 3-noted whistle—lowest, high, lower.

302 WHITEBROWED YUHINA *Yuhina castaniceps* (Moore)   p. 97
Size Sparrow − ; length 13 cm (5 in.).

Field Characters *Above*, crown and crest grey scalloped with paler grey; rufous-brown ear-coverts and a narrow white supercilium. Back and wings grey-brown. Tail dark brown, rounded, when spread showing white tips of outer rectrices. *Below*, greyish white (subspecies *rufigenis* and *plumbeiceps*). Sexes alike.

Status, Habitat, etc. Resident, fairly common. Foothills from 600 to 1500 m or so: secondary forest with scrubby undergrowth. Parties of 20 to 30 among the roving mixed flocks. Habits as of mesias, qq.v. (287–8). Food: chiefly insects; also seeds. Call: loud cheeping or twittering contact notes *chir-chit . . . chir-chit . . .*

303 WHITENAPED YUHINA *Yuhina bakeri* Rothschild   p. 113
Size Sparrow − ; length 13 cm (5 in.).

Field Characters A perky hair-brown tit-like bird with upstanding chestnut crest and prominent white nape-patch. *Above*, chiefly rusty brown and olive-brown with faint white shaft-streaks. *Below*, throat white; breast vinaceous, finely dark streaked; belly olivaceous; vent ferruginous. Sexes alike.

Status, Habitat, etc. Common resident, subject to vertical movements. Duars, foothills and up to 2000 in winter, 3000 m in summer: evergreen forest and secondary jungle. Arboreal. Parties in the mixed foraging flocks of other small babblers etc. Food: mainly insects; also berries. Call: a shrill *chip* and a soft chattering.

304 YELLOWNAPED YUHINA *Yuhina flavicollis* Hodgson   p. 97
Size Sparrow − ; length 13 cm (5 in.).

Field Characters Tit-like profile, erectile chocolate-brown crest, rusty yellow nape and white eye-ring diagnostic. Confusable with *Y. occipitalis* (306) but which has a grey crest and bright rufous nape (subspecies *flavicollis* and *rouxi*). Sexes alike.

Status, Habitat, etc. Common resident, from 1800 to 3000 m in summer, descending lower and into the duars in winter: deciduous forest and secondary jungle. Arboreal. Habits, behaviour and ecology as of other yuhinas, mesias and suchlike small sociable babblers. Food: insects, berries, flower nectar. Call: a murmuring conversational twitter interspersed by a harsh *chi-chi-chiu*. Song, of pleasant warbles and querulous screeches.

305 STRIPETHROATED YUHINA *Yuhina gularis* Hodgson   p. 97
Size Sparrow; length 14 cm (5½ in.).
Field Characters A plump, active brown tit-like bird with erectile brown crest, striped throat, and an orange-fulvous longitudinal bar on blackish wings. *Above*, chiefly olive-brown. *Below*, vinaceous and tawny brown with dark brown streaking on throat (subspecies *gularis*). Sexes alike.
Status, Habitat, etc. Common resident between 2400 and 3600 m, descending lower in winter, rarely even to the duars: mixed conifer and rhododendron forest, occasionally low scrub or bamboo. Habits, behaviour, food, etc. as of mesias and other similar small babblers. Call: a distinctive long-drawn-out *kweeeeee*: a continual quiet, rustling *shr . . . shr . . .* while foraging.

306 SLATYHEADED YUHINA *Yuhina occipitalis* Hodgson   p. 113
Size Sparrow; length 13 cm (5 in.).
Field Characters *Above*, chiefly olive-brown: head grey; an erectile crest grey in front, bright rufous posteriorly; a conspicuous pale eye-ring and a black malar stripe. *Below*, throat and breast vinaceous; rest pale rufous (subspecies *occipitalis*). Sexes alike. The somewhat similar *Y. flavicollis* (304) has a chocolate-brown crest and rusty yellow nape.
Status, Habitat, etc. Common resident between 2400 and 3900 m, down to 1500 m in winter: evergreen rhododendron and oak forest. Arboreal and sociable. Behaviour and ecology as of other yuhinas. Food: insects, berries, flower nectar. Call: deep conversational churring notes.

307 BLACKCHINNED YUHINA *Yuhina nigrimenta* Hodgson   p. 113
Size Sparrow − ; length 11 cm (4½ in.).
Field Characters A small babbler with erectile black crest, black lores and chin, and black-and-red bill. *Above*, crest black with scale-like grey edging to the feathers; nape and sides of head grey. *Below*, chin black, throat white, rest pale fulvous (subspecies *nigrimenta*). Sexes alike.
Status, Habitat, etc. Fairly common resident. Chiefly lower foothills and up to 1800 m: evergreen forest and secondary jungle, especially overgrown clearings etc. Very gregarious, active, restless and noisy. Flocks of up to 20 or more among the mixed foraging parties. Habits and behaviour as of other yuhinas and small babblers. Food: insects,

berries, seeds, flower nectar. Call: a lively chorus of low cheeping twitters while feeding; occasional louder and shriller notes.

308 **WHITEBELLIED YUHINA** *Yuhina zantholeuca* (Hodgson) p. 113
Size Sparrow − ; length 11 cm (4½ in.).
Field Characters A small slightly tufted babbler, olive-green above, greyish white below with yellow under tail-coverts (subspecies *zantholeuca*). Sexes alike.
Status, Habitat, etc. Resident, locally common. Duars, foothills and up to at least 2500 m: evergreen and moist-deciduous secondary forest. Less gregarious than other yuhinas, but like them usually found among the mixed itinerant foraging flocks. Sprightly, restless behaviour reminiscent of a leaf warbler. **Food**: insects, berries, flower nectar. **Call**: unrecorded. A very silent species.

309 **GOLDENBREASTED TIT-BABBLER** *Alcippe chrysotis* (Blyth)
p. 97
Size Sparrow − ; length 11 cm (4½ in.).
Field Characters A small brightly coloured babbler with yellow underparts and no supercilium. *Above*, crown blackish; ear-coverts silver-grey; back olive. Wings blackish, with orange-yellow outer edge and an orange longitudinal patch; secondaries tipped with white. Tail brown, the basal two-thirds edged with orange-yellow. *Below*, throat grey with silvery stippling; rest yellow (subspecies *chrysotis*). Sexes alike.
Status, Habitat, etc. Resident, rather scarce, between 2400 and 3000 m, descending lower in winter: dense growth on hillsides, particularly bamboo jungle. Large parties of up to 20 birds or more, usually associated with mixed foraging flocks of tinies. Actions very tit-like. **Food**: insects, berries, seeds. **Call**: a continual low conversational twitter.

310 **DUSKY GREEN TIT-BABBLER** *Alcippe cinerea* (Blyth) p. 113
Size Sparrow − ; length 10 cm (4 in.).
Field Characters A small greyish olive-and-yellow babbler with a well-marked yellow supercilium. *Above*, crown and nape yellowish green with black scaly markings. A black stripe along side of crown above supercilium and another through eye below supercilium. *Below*, yellow; olivaceous on belly. Sexes alike.
Status, Habitat, etc. Resident, locally distributed. Foothills and up to 2100 m: deep evergreen forest glades and overgrown clearings. Habits, food and ecology as of yuhinas and other small babblers. **Call**: a low *chip-chip* and a soft conversational twittering.

311 CHESTNUTHEADED TIT-BABBLER

Alcippe castaneceps (Hodgson)   p. 97

Size Sparrow − ; length 10 cm (4 in.).

Field Characters *Above*, olive; forehead, crown and nape chestnut, streaked with white and rufous; a broad white supercilium and blackish post-ocular stripe. Ear-coverts mostly white; a narrow dark malar stripe. Wing with a black and rufous shoulder-patch. *Below*, whitish, with olive-rufous flanks (subspecies *castaneceps*). Sexes alike.

Status, Habitat, etc. Common resident. Breeds between 1500 and 3000 m; descends lower in winter: heavy evergreen undergrowth at forest edge and on abandoned clearings. Active flocks by themselves or among the mixed foraging assemblages of kindred species. Food: chiefly insects. Call: a distinctive three-noted *tu-twee-twee*, rising in pitch.

312 WHITEBROWED TIT-BABBLER *Alcippe vinipectus* (Hodgson)

p. 97

Size Sparrow − ; length 11 cm (4½ in.).

Field Characters A fluffy tit-like high-elevation babbler. *Above*, crown, ear-coverts and back brown; a broad white supercilium from eye to nape; a dark brown stripe above this. Rump and wings rusty, the latter (when closed) with a black line and pale outer edge. *Below*, throat and breast white streaked with brown; lower belly olive-brown (subspecies *chumbiensis*). Sexes alike.

Status, Habitat, etc. Common resident, subject to vertical movements— 2400 to 4200 m in summer, between 1500 and 3000 m in winter. Affects rhododendron and juniper scrub, undergrowth in forest clearings and edges, especially ringal bamboo. Flocks of up to 20 or so foraging in bushes and low trees, by themselves or in the mixed itinerant assemblages. Actions characteristic of minlas and other small babblers. Confiding, fussy and inquisitive. Food: insects, berries, seeds. Call: a soft, high-pitched and incessant *chip, chip* . . .

313 BROWNHEADED TIT-BABBLER *Alcippe cinereiceps* (Verreaux)

p. 113

Size Sparrow − ; length 11 cm (4½ in.).

Field Characters. *Above*, head chocolate-brown, no white supercilium; sides of head, and nape, reddish brown. Rest of upper- and underparts as in Whitebrowed Tit-Babbler (312) (subspecies *ludlowi*). Sexes alike.

Status, Habitat, etc. Common resident, subject to some vertical movement: up to 3500 m elevation in summer, 2200 or so in winter. Affects secondary scrub jungle—brambles, bamboo, etc. Parties of 6 to 10, often among the mixed foraging associations. Habits, behaviour and food as in 312. Call: a tit-like *cheep*. A rattling song of 3 or 4 notes.

314 **REDTHROATED TIT-BABBLER** *Alcippe rufogularis* (Mandelli)

p. 113

Size Sparrow − ; length 12 cm (5 in.).

Field Characters *Above,* crown dark rufous-brown, bordered by a broad black stripe from forehead to nape; a white supercilium and eye-ring. Rest of upperparts rufescent brown. *Below,* chin and throat white with a broad chestnut band; belly whitish, flanks olive-brown (subspecies *collaris*). Sexes alike.

Status, Habitat, etc. Common resident, chiefly in Arunachal Pradesh, from plains level to 900 m: bamboo jungle and secondary undergrowth in evergreen forest. Small restless parties, often mixed with other small babblers, skulking in low bushes, often feeding on the ground. Food: mainly insects. Call: a rather musical cheeping *chree-chree* while on the move.

315 **QUAKER BABBLER** *Alcippe nipalensis* (Hodgson)   p. 97

Size Sparrow − ; length 12 cm (5 in.).

Field Characters *Above,* head grey; a blackish supercilium from eye to nape; a conspicuous white eye-ring or 'spectacles' (as in White-eye). Rest of upperparts fulvous brown. *Below,* uniformly buff (subspecies *nipalensis* and *commoda*). Sexes alike.

Status, Habitat, etc. Common resident, subject to vertical movements, from the lower foothills up to 1800 m in winter, to 2400 m in summer. Affects undergrowth in moist-deciduous or evergreen forest, and bamboo jungle. Active, restless flocks associated with other small babblers, searching the foliage in acrobatic tit-like postures, and flowing from tree to tree. Food: insects, berries, flower nectar. Call: a constant conversational twittering—a shrill whinnying note.

316 **CHESTNUTBACKED SIBIA** *Heterophasia annectens* (Blyth)

p. 81

Size Bulbul; length 18 cm (7 in.).

Field Characters *Above,* crown, hindneck and upper back black; hindneck streaked with white; centre of back and rump chestnut. Wings black with a chestnut bar; primaries edged ashy, tertials tipped white. *Below,* white; flanks and vent fulvous (subspecies *annectens*). Sexes alike.

Status, Habitat, etc. Resident, rather scarce, subject to vertical movements. Between 1200 and 2300 m in summer, down to the foothills in winter: dense humid evergreen forest. Arboreal. Small parties keeping chiefly to the canopy foliage. Creeps or clambers along branches like nuthatch, probing for insects among the moss and lichen, and in bark crevices. Food: insects and seeds. Call: a clear single whistle. Song, a 4-noted musical phrase. Alarm-note *chirr-r-r*.

317 **BLACKCAPPED SIBIA** *Heterophasia capistrata* (Vigors)   p. 128
Size Bulbul + ; length 20 cm (8 in.).
Field Characters General effect bulbul-like. *Above*, crown and erectile crest black. Upper back and rump rufous; middle back sooty brown tinged with grey. Wings bluish slaty with a black shoulder-patch; a white patch conspicuous in flight. Tail long, graduated, rufous with broad black-and-grey terminal band. *Below*, entirely cinnamon colour (subspecies *bayleyi*). Sexes alike.
Status, Habitat, etc. Common resident, subject to vertical movements, from 1800 to *c.* 2500 m in summer, descending to the foothills in winter: tall moist-deciduous and evergreen forest. Arboreal. Pairs or small parties. Very active and lively. Hops swiftly up and along moss-covered boughs and peers into crevices and under leaves in quest of insects. Food: insects, berries, flower nectar. Call: a loud *tee-riri-reeri-reeri* like the jingle of a silver bell, frequently repeated. A clear flute-like 6-noted song.

318 **BEAUTIFUL SIBIA** *Heterophasia pulchella* (Godwin-Austen)
Size Bulbul + ; length 22 cm (8½ in.).
Field Characters *Above*, head bluish slate; forehead and a line through eye black; back and rump slaty. Wings with pale bluish outer edge and a conspicuous black shoulder-patch. Centre of tail dark brown with black edges, black subterminal and slaty terminal band. *Below*, uniform pale slate-grey. Sexes alike.
Status, Habitat, etc. Resident, subject to vertical movements. Between 2100 and 3000 m in summer, from 1200 to 2700 m in winter (Arunachal Pradesh): mossy forest. Arboreal. Pairs or small parties in tall trees hopping actively along the trunk and boughs searching for insects under moss and bark. Food: insects, seeds and vegetable matter. Call: more or less as in 317.

319 **LONGTAILED SIBIA** *Heterophasia picaoides* (Hodgson)   p. 128
Size Bulbul, with a long tail. Overall length 30 cm (12 in.).
Field Characters *Above*, dark slaty grey. Wings blackish with a white patch. Tail very long, graduated, tipped with whitish. *Below*, uniform pale grey. Under surface of tail looks barred black-and-grey (subspecies *picaoides*). Sexes alike.
Status, Habitat, etc. Resident, locally common, with some seasonal altitudinal movements—from the base of the hills up to 900 m in winter, a thousand metres or so higher in summer: evergreen forest and clearings with tall trees. Pairs or parties, sometimes of 30 or 40, keeping mostly to tree tops, drifting one after another from tree to tree while foraging,

shooting down like arrows into lower growth when alarmed. Food: insects, berries, flower buds, nectar. Call: a rapid high-pitched *tsip-tsip-tsip-tsip* ... Song, a rich whistling 6-noted phrase of thrush quality ending in *wheet-whew*.

FLYCATCHERS: Muscicapinae

320 **SOOTY FLYCATCHER** *Muscicapa sibirica* Gmelin   p. 128
Size Sparrow − ; length 13 cm (5 in.).
Field Characters *Above*, dark grey-brown with strikingly large eyes and a pale eye-ring. *Below*, grey-brown with a whitish throat-patch and centre of belly (subspecies *cacabata*). Sexes alike.
Status, Habitat, etc. Altitudinal migrant, locally common; breeds between 2400 and 3500 m, descends lower in winter: open conifer and oak forest. Keeps singly, but 3 or 4 often seen hawking in a restricted glen. Launches agile twisting and turning aerial sorties after flies and circles back to the perch after each. Hunts like this all day and often far into evening dusk. Food: flies, gnats and suchlike insects. Call: normally silent. Song, high-pitched, thin, reedy, usually of 3 descending notes *tsee-see-see*.

321 **BROWN FLYCATCHER** *Muscicapa latirostris* Raffles
Size Sparrow ± ; length 14 cm (5½ in.).
Field Characters An ashy brown flycatcher with strikingly large eyes and a conspicuous white ring round them. *Below*, sullied white, greyish on breast and flanks, faintly streaked with olive-brown. In poor light, the glistening white throat is prominent. Sexes alike. Sooty Flycatcher (320) is darker below and has a much smaller throat-patch.
Status, Habitat, etc. Summer visitor (breeding unconfirmed) to Jalpaiguri duars and Bhutan foothills: open mixed deciduous forest, overgrown nullahs, etc. preferably neighbourhood of streams. Usually solitary. Typical flycatcher in habits and food (see 320). Call: a feeble *chi-chir-ri-ri-ri*. Song, short, of a single plaintive note slowly but freely uttered.

322 **FERRUGINOUS FLYCATCHER**
    Muscicapa ferruginea (Hodgson)
Size Sparrow − ; length 13 cm (5 in.).
Field Characters A small brown flycatcher with rusty belly. *Above*, head blackish with a conspicuous pale eye-ring. Back rusty brown and ferruginous. Wings blackish, the tertials prominently edged with pale rufous. Tail chestnut. *Below*, chin, throat and centre of belly white;

Plate 28, artist Winston Creado

FLYCATCHER-WARBLERS, LEAF WARBLERS

1 R   **ALLIED FLYCATCHER-WARBLER,** *Seicercus affinis*   page 186
Sparrow – . Whitish eye-ring. Black lateral coronal bands. Up to 2300 m.

2 R   **GREYHEADED FLYCATCHER-WARBLER,** *Seicercus xanthoschistos*   187
Sparrow – . Outer rectrices largely white. White supercilium. Up to 2700 m.

3 M   **TICKELL'S LEAF WARBLER,** *Phylloscopus affinis*   182
Sparrow – . No wing-bar. Yellow supercilium and underparts. Up to 2100 m (winter).

4 R   **BROADBILLED FLYCATCHER-WARBLER,** *Abroscopus hodgsoni*   189
Sparrow – . Forehead chestnut. Rump yellow. Throat and breast grey. 1100–2700 m.

5 R   **YELLOWBELLIED FLYCATCHER-WARBLER,** *Abroscopus superciliaris*   188
Sparrow – . Forehead grey. Broad white supercilium. Up to 900 m.

6 R   **GREYCHEEKED FLYCATCHER-WARBLER,** *Seicercus poliogenys*   187
Sparrow – . Head slaty; 2 black lateral bands. White eye-ring. Up to 3000 m.

7 M   **YELLOWBROWED LEAF WARBLER,** *Phylloscopus inornatus*   183
Sparrow – . Two wing-bars. Yellowish supercilium. Two dark lateral, one buffish median, coronal bands. Up to 1800 m.

8 R   **PALLAS'S LEAF WARBLER,** *Phylloscopus proregulus*   184
Sparrow – . Two wing-bars. Yellow rump-band. Coronal bands. Up to 4200 m.

9 R   **GREYFACED LEAF WARBLER,** *Phylloscopus maculipennis*   184
Sparrow – . Two wing-bars. Yellow rump. Whitish median coronal stripe. Grey head and throat. White supercilium. Up to 3400 m.

10 M   **YELLOWFACED LEAF WARBLER,** *Phylloscopus cantator*   186
Sparrow – . Two wing-bars. Yellow supercilium. Throat and vent bright yellow. Duars and foothills.

11 R   **SMOKY LEAF WARBLER,** *Phylloscopus fuligiventer*   182
Sparrow – . Greenish yellow supercilium. Up to 4300 m.

12 R   **LARGEBILLED LEAF WARBLER,** *Phylloscopus magnirostris*   185
Sparrow – . Single faint wing-bar. Yellow supercilium. Up to 3600 m.

13 R   **BLYTH'S CROWNED LEAF WARBLER,** *Phylloscopus reguloides*   186
Sparrow – . Under tail-coverts whitish. Up to 3500 m.

14 M   **CROWNED LEAF WARBLER,** *Phylloscopus occipitalis*   185
Sparrow – . Like 13, but under tail-coverts yellow. Up to 2000 m.

0 2 4 6 cm

Winston Creado

JPIrani 1972

0 2 4 6 8 10 cm

Plate 29, artist J. P. Irani

REDSTARTS, SHORTWINGS, SHAMA,
ROBINS, BLUE CHAT

1 R **PLUMBEOUS REDSTART,** *Rhyacornis fuliginosus* page 200
Sparrow – . ♂ Tail chestnut. ♀ Tail white with brown terminal band.
Up to 4000 m.

2 R **LESSER SHORTWING,** *Brachypteryx leucophrys* 190
Sparrow – . ♂ Underparts white; a grey breast-band. ♀ Brown above,
white below. Brown breast-band. Foothills up to 3000 m.

3 R **WHITEBELLIED REDSTART,** *Hodgsonius phoenicuroides* 200
Bulbul. ♂ Rufous patch on rectrices. ♀ Brown with similar tail-
patches. Foothills up to 3900 m.

4 R **SHAMA,** *Copsychus malabaricus* 196
Bulbul. White rump. Long black-and-white graduated tail. Up to
500 m.

5 R **BLUEFRONTED REDSTART,** *Phoenicurus frontalis* 198
Sparrow ±. ♀ Brown. Tail in both sexes rufous with blackish
terminal band. 1000–4500 m.

6 R **WHITETAILED BLUE ROBIN,** *Cinclidium leucurum* 200
Bulbul – . ♀ Brown. Tail (both sexes) with white patches near base.
Duars up to 2700 m.

7 R **RUSTYBELLIED SHORTWING,** *Brachypteryx hyperythra* 190
Sparrow – . ♀ Brown. Short white supercilium both sexes. 1100–
2900 m.

8 R **DAURIAN REDSTART,** *Phoenicurus auroreus* 199
Sparrow ±. ♂ Middle of back black. White wing-patch (secondaries). ♀ Brown, with similar wing-patch. Foothills up to 3700 m.

9 M **HODGSON'S REDSTART,** *Phoenicurus hodgsoni* 197
Sparrow ±. ♂ Whitish forecrown and supercilium. White wing-
patch. ♀ Brown. Grey breast. No wing-patch. Duars up to 2800 m.

10 R **BLACK REDSTART,** *Phoenicurus ochruros* 197
Sparrow ±. ♀ Brown. Tail rufous. Duars up to 5200 m.

11 R **RUFOUSBELLIED BUSH ROBIN,** *Erithacus hyperythrus* 195
Sparrow – . ♂ Supercilium and rump bright blue. Vent white.
♀ Brown. Rump slaty blue. Vent white. Foothills up to 3800 m.

12 R **WHITEBROWED SHORTWING,** *Brachypteryx montana* 191
Sparrow – . ♂ Long white supercilium. ♀ Brown above, paler below.
Vent rufescent. 300–3300 m.

13 R **REDFLANKED BUSH ROBIN,** *Erithacus cyamurus* 194
Sparrow ±. ♀ Brown. Rump and tail bluish. Flanks orange. Foothills
up to 4400 m.

14 R **BLUE CHAT,** *Erithacus brunneus* 194
Sparrow ±. ♀ Olive-brown above, ochraceous white below. Tail
rufous. 1600–3300 m.

breast dark-spotted olive-brown; rest pale rusty. Sexes alike. Brooks's, *M. poliogenys* (336), the only other brown flycatcher with rusty belly, has olive-brown back and tail, and no rufous edges in wing.

Status, Habitat, etc. Probably only summer visitor, from 1800 to 3300 m; winter status unclear. Affects fir and oak forest. Usually solitary, quiet, retiring, and rather crepuscular. Typical flycatcher in habits and food. Call: unrecorded.

323 REDBREASTED FLYCATCHER *Muscicapa parva* Bechstein
p. 128

Size Sparrow — ; length 13 cm (5 in).

Field Characters Male. *Above*, pale brown. Tail black with a white patch on each side of basal half, conspicuous when characteristically cocked and flicked upwards. *Below*, chin and throat orange-rufous; breast grey, rest white. Female and subadult male have whitish throat and greyish breast (subspecies *albicilla*).

Status, Habitat, etc. Winter visitor, common. Duars and foothills to 1500 m: groves, orchards and scrub jungle. Usually solitary. Typical flycatcher in habits and food. Frequently descends to ground to pick up an insect with smart upward flicks of cocked tail and flits back to perch in overhanging branch. Call: a double *tick-tick* when jerking tail and twitching wings; also a plaintive *phwee-phwee-phwee* or *weeit-weeit-weeit* ...

324 ORANGEGORGETED FLYCATCHER
Muscicapa strophiata (Hodgson)
p. 129

Size Sparrow — ; length 13 cm (5 in.).

Field Characters Male. *Above*, forehead white; upperparts olive-brown. Tail black, white on each side of basal half. *Below*, chin and upper throat black; an orange-rufous patch on lower throat bounded below and on sides by dark slaty. Belly ashy, paling to white on vent; flanks olive-brown (subspecies *strophiata*). Female paler, with smaller orange gorget, and ashy throat; sometimes like male. Similar tail pattern to Redbreasted misleading, but white forehead always diagnostic.

Status, Habitat, etc. Resident, moving altitudinally. Breeds between 2500 and 3000 m; winters lower and in the foothills. Oak, rhododendron and conifer forest in summer; shady forest edges and clearings in winter. Solitary or pairs. Typical flycatcher in habits and food; particularly resembles Redbreasted (323), q.v. Call: a low *tik-tik*. Alarm, a croaking *churr*. Song, a short, spirited, triple-noted *tin-ti-ti*.

325 WHITEGORGETED FLYCATCHER
Muscicapa monileger (Hodgson)

Size Sparrow − ; length 11 cm (4½ in.).

Field Characters A short-tailed brown flycatcher with a black-bordered white gorget. *Above*, olive-brown; short broad fulvous supercilia nearly meeting on forehead. Tail ferruginous. *Below*, olive-brown with a triangular white throat-patch bordered with black; centre of belly whitish (subspecies *monileger* and *leucops*). Sexes alike.

Status, Habitat, etc. Resident, uncommon; from the foothills up to 2000 m: dense bush jungle, scrub-covered ravines and undergrowth in tropical evergreen forest. Solitary and retiring. A typical flycatcher in habits and food. Call: some chattering notes; a weak but pleasant little song.

326 RUFOUSBREASTED BLUE FLYCATCHER
Muscicapa hyperythra Blyth   p. 128

Size Sparrow − ; length 11 cm (4½ in.).

Field Characters Male. *Above*, slaty blue with a conspicuous white supercilium; forehead and cheeks black. Tail blue-black above. *Below*, chin black; rest orange-rufous. Confusable with Blue Chat (399) which is larger and has longer supercilium; also with very similar Rustybellied Shortwing (394) which has the supercilium less prominent, and longer legs. Female. *Above*, olive-brown; supercilium and eye-ring fulvous. *Below*, orange-rufous, duller than in male, paler on throat and belly.

Status, Habitat, etc. Resident, subject to vertical movements. Breeds between 2000 and 3000 m; winters in the foothills and duars: dense primary forest with luxuriant undergrowth. Singly or pairs, usually in low bushes and dank herbage. Flits among branches or runs mouse-like among fallen tree-stems and débris flicking tail and making short aerial sallies after midges. On ground easily mistaken for Shortwing. Food: insects. Call: a short sweet percussive song of 4 notes.

327 RUSTYBREASTED BLUE FLYCATCHER
Muscicapa hodgsonii (Verreaux)

Size Sparrow − ; length 13 cm (5 in.).

Field Characters Male. *Above*, slaty blue; lores and cheeks velvety black. Tail blackish, with bases of all rectrices white except central pair. *Below*, throat, breast and flanks orange-rufous; lower belly and vent buffish. Female. *Above*, olive-brown; a pale eye-ring. *Below*, olive-buff, whitish on belly. Confusable with female *M. parva* (323) but the latter has a whitish throat-patch and black tail with white bases instead of plain blackish.

Status, Habitat, etc. Altitudinal migrant, locally common. Breeds between 2100 and 3900 m; winters in the foothills and duars. Pine or fir forest in summer; dense scrub and mixed bamboo and tree forest in winter. A quiet shy flycatcher doing most of its hunting from the canopy of trees. **Food**: insects. **Call**: song of clear, pleasant robin-like rippling whistles, constantly uttered.

328 LITTLE PIED FLYCATCHER *Muscicapa westermanni* (Sharpe)

p. 128

Size Sparrow − ; length 10 cm (4 in.).

Field Characters A diminutive black-and-white flycatcher. *Above*, black, with a broad white supercilium from lores to nape, a large white wing-patch, and white sides to base of tail. *Below*, white (subspecies *collini* and *australorientis*). **Female**. *Above*, olive-brown with a pale wing-bar; upper tail-coverts bright rufous-brown. *Below*, throat white; rest smoky white.

    A passable miniature of *M. ruficauda* of western Himalayas, and confusable on a casual sighting also with several other brown female flycatchers.

Status, Habitat, etc.   Altitudinal and short-range migrant; not common. Breeds between 1200 and 2500 m; winters in the foothills, duars and adjacent plains. Steep hillsides in deciduous and evergreen forest in summer; secondary growth, especially along streams, in winter. Singly or small parties, often in association with other small insectivorous birds, mostly in crowns of trees. Feeding habits and food typically flycatcher. **Call**: a single mellow *tweet*. Song, a thin high-pitched *pi-pi-pi-pi* followed by a low rattle *churr-r-r-r-r*.

329 LITTLE BLUE-AND-WHITE FLYCATCHER
Muscicapa superciliaris Jerdon

Size Sparrow − ; length 10 cm (4 in.).

Field Characters **Male**. *Above*, deep blue. *Below*, sides of head and neck deep blue; centre of throat and breast, and whole belly, white, especially glistening on throat (subspecies *aestigma*). **Female**. *Above*, mouse-grey; tail blackish, edged with blue. *Below*, sides of neck and breast greyish white; centre of throat and breast, and whole belly, glistening white. Confusable on a casual sighting with rather similar females of *M. westermanni* (328) and *M. leucomelanura* (330).

Status, Habitat, etc.   Uncommon summer visitor, between 2000 and 3000 m; at lower elevations and in the duars in winter. Fairly open forest of oak, rhododendron, etc. in summer; various types of woodland in winter. Singly or pairs, often among the mixed hunting parties of tinies in the foliage canopy. A typical flycatcher in habits and food. **Call**: a soft, repeated *tik*. Alarm, a soft *trrr*. Song, an oft-repeated *che-chi-purr*.

330 **SLATY BLUE FLYCATCHER**
Muscicapa leucomelanura (Hodgson)
p. 129
Size Sparrow − ; length 10 cm (4 in.).
Field Characters Male. *Above*, slaty blue, brighter on forehead. Tail black, conspicuously white on each side of base. *Below*, sides of head and throat blackish slate; throat white; rest of underparts white tinged with rufous or olive (subspecies *minuta*). Female. *Above*, olive-brown; tail rufous. *Below*, buffish. Distinguished from very similar females of *M. westermanni* (328) and *M. superciliaris* (329) by its rufous tail. *M. ruficauda* of western Himalayas is larger and does not flick its rufous tail.
Status, Habitat, etc. Altitudinal migrant, locally common in summer between 2700 and 4000 m; winters in the foothills down to the duars. Oak, rhododendron and pine forest in summer; thickets of reeds and undergrowth in winter. Solitary or pairs. Restless and rather secretive, keeping to low vegetation. Perches with wings drooping and constantly flicks tail upwards, the male thus flashing the white lateral patches. Food: insects. Call: a rapid *ee-tik-tik-tik-tik* accompanied by tail-flicking.

331 **SAPPHIREHEADED FLYCATCHER**
Muscicapa sapphira (Blyth)
p. 129
Size Sparrow − ; length 12 cm (5 in.).
Field Characters Male. *Above*, forehead, crown, rump and tail bright ultramarine blue; a black line through eye; sides of head and back deep purplish blue. *Below*, chin, throat and upper breast orange-rufous; an interrupted breast-band deep blue. Belly ashy. Female. *Above*, rufescent olive-brown; rump rufous-brown; an ochraceous eye-ring. *Below*, chin, throat and breast orange-rufous and olive-brown; belly whitish. Confusable with female *M. hyperythra* (326) but which has a short fulvous supercilium.
    Young male (illustrated) like female but with wings, tail and rump purplish blue as in adult male, and no breast-band.
Status, Habitat, etc. Altitudinal migrant, fairly common. Summer limit little known; winters from 1700 down: evergreen forest. Singly or pairs in high undergrowth or low branches of trees. Hunts mostly in foliage canopy. Typical flycatcher habits and food. Call: little recorded except *tik-tik* accompanied by tail flicks.

332 **LARGE NILTAVA** *Muscicapa grandis* (Blyth)
p. 129
Size Bulbul; length 20 cm (8 in.).
Field Characters Male. A large, rather sluggish flycatcher, deep purplish

blue and black, with crown, rump, and a patch on each side of neck brilliant cobalt blue (subspecies *grandis*). Female. Overall olive-brown and fulvous-brown with a paler (fulvous) throat and a distinctive pale blue patch on each side of neck.

Status, Habitat, etc. Resident, fairly common, subject to vertical movements. Breeds between 1800 and 2700 m; winters from 2000 down to the duars: dense humid forest and secondary jungle especially on hillsides and near streams. Singly or pairs. Flits among low bushes; more often feeds on ground than in typical flycatcher fashion. Food: insects and berries. Call: song, a rather mournful ascending whistle of 3 or 4 notes ending interrogatively, *whee-whee-wip?*

333 SMALL NILTAVA *Muscicapa macgrigoriae* (Burton)   p. 128

Size Sparrow – ; length 11 cm (4½ in.).

Field Characters Male. *Above*, forecrown, rump, and a patch on each side of neck brilliant ultramarine blue; forehead and lores black; rest of upperparts deep purplish blue. *Below*, throat deep purplish blue; rest ashy grey, whitish on belly (subspecies *signata*). Female. *Above*, rufescent olive-brown. Wings and tail rusty brown. *Below*, throat fulvous; a pale blue patch on each side of neck. Rest of underparts fulvous olive-brown. Blue patches on each side of neck in both sexes is the hall-mark of all niltavas.

Status, Habitat, etc. Common resident, subject to vertical movements. Breeds from 900 to at least 2000 m; winters from 1400 m down to the duars and adjacent plains: shady glades and clearings in secondary evergreen forest, usually along streams. Usually solitary. Sprightly and active, with typical flycatcher habits and food. Call: a high-pitched *see-see*, the second note lower. Song, a very high-pitched thin rising and falling whistle, *twee-twee-ee-twee*.

334 RUFOUSBELLIED NILTAVA *Muscicapa sundara* (Hodgson)
p. 128

Size Sparrow; length 15 cm (6 in.).

Field Characters Male. *Above*, forehead black; crown, rump, shoulders, and a patch on each side of neck ultramarine blue; sides of head, and back, purplish blue-black. *Below*, throat black; rest orange-rufous (subspecies *sundara*). Confusable with very similar *M. vivida* (335) which lacks the blue neck-patches—the niltava hall-mark. Female. *Above*, olive-brown, ochraceous on rump; a pale eye-ring; a blue patch on each side of neck. Tail rusty brown. *Below*, chin and upper throat fulvous-olive; lower throat white; rest olive-brown.

Status, Habitat, etc. Common resident, subject to vertical movements.

Breeds between 1800 and 3200 m; winters down to the duars and adjacent plains: dense secondary undergrowth and brush-covered hillsides. Usually solitary; in winter often among itinerant hunting flocks. Habits partly flycatcher, partly reminiscent of Blue Chat (399). Food: insects and berries. Call: alarm-note, a harsh, scolding *tr-r-r-tchik*. A high-pitched *tzi, tzi, tzi*, and a squeaky, grating song.

335 RUFOUSBELLIED BLUE FLYCATCHER

Muscicapa vivida (Swinhoe)

Size Sparrow + ; length 18 cm (7 in.).

Field Characters Male. Very similar to Rufousbellied Niltava (334) but lacks the distinctive blue patch on sides of neck (subspecies *oatesi*). Female. *Above*, dark olive-brown; crown grey-brown; a fulvous eye-ring. *Below*, throat fulvous; rest olive-brown. Distinguished from similar females of Large and Rufousbellied Niltavas by absence of the diagnostic blue patch on sides of neck.

Status, Habitat, etc. Rare. Recorded in summer between 2100 and 2700 m (Sikkim, Arunachal): dense brushwood in evergreen forest. Singly or pairs. Behaviour, food, etc. as in Large Niltava (332). Call: a clear whistle.

336 BROOKS'S FLYCATCHER *Muscicapa poliogenys* (Brooks) p. 128

Size Sparrow − ; length 14 cm (5½ in.).

Field Characters A rather nondescript species, confusable with several brown female flycatchers with more or less rufous underparts. *Above*, olive-brown, greyer on crown and sides of head; a pale eye-ring. Tail rufous-brown. *Below*, throat buff; rest of underparts ochraceous, darker on breast (subspecies *poliogenys* and *cachariensis*). Sexes alike.

Status, Habitat, etc. Common resident, from the edge of the plains up to 1500 m: deciduous and evergreen forest; more open jungle in winter. Singles or pairs. Habits and behaviour part flycatcher, part chat. Food: insects. Call: alarm-note, *tik-tik-tik-tik*. Song, a pleasing mellow trill, variously rendered.

337 PALE BLUE FLYCATCHER *Muscicapa unicolor* (Blyth) p. 128

Size Sparrow ± ; length 16 cm (6 in.).

Field Characters Male. *Above*, blue; brighter on forehead, supercilium and shoulder, deeper on tail. *Below*, throat pale blue; breast blue fading to whitish on belly. Under tail-coverts scalloped grey and white. Confusable only with Verditer Flycatcher (340) but which is more blue-green and lacks whitish belly. Female. *Above*, olive-brown; wing browner; tail more reddish brown. A pale eye-ring. *Below*, pale grey-brown. May be confused with females of both *M. hodgsonii* (327) and *M. vivida* (335).

Status, Habitat, etc. Status imperfectly known. From the foothills in winter up to at least 1800 m in summer, breeding around 1500 m: dense humid secondary forest and bamboo jungle on hillsides. Solitary or pairs, in undergrowth as well as high trees. Food: insects. Call: the characteristic *trr-r-r* of flycatchers, accompanied by upward jerks of the half-cocked tail. A rich, beautiful song.

338 BLUETHROATED FLYCATCHER
Muscicapa rubeculoides (Vigors)   p. 128

Size Sparrow − ; length 14 cm (5½ in.).

Field Characters Male. *Above*, dull ultramarine blue, brighter on forehead and supercilium. Lores black. *Below*, throat dark blue; breast rufous; belly and vent buffy white (subspecies *rubeculoides*). Combination of dark blue throat, rufous breast and white belly diagnostic. Female. *Above*, olive-brown; rump tinged rufous. A pale eye-ring. *Below*, throat buff; breast ochraceous; rest white. Casually confusable with several brown female flycatchers with rufous in underparts.

Status, Habitat, etc. Partial migrant, some still on breeding ground in winter. Locally up to 2000 m in summer; down to the duars and plains level in winter: forest undergrowth, well-wooded nullahs and gardens, and bamboo jungle. Singly or pairs in bushes and low trees. Typical flycatcher habits and food. Call: characteristic flycatcher *clik*, *clik*. Alarm-note, *chr-r*, *chr-r*. Song a clear metallic trill of same pattern as of peninsular Tickell's, only somewhat richer.

339 LARGEBILLED BLUE FLYCATCHER
Muscicapa banyumas Horsfield

Size Sparrow − ; length 14 cm (5½ in.).

Field Characters Male. *Above*, indigo-blue, brighter on forehead and shoulders; lores black. *Below*, throat, breast and flanks orange-rufous; belly and vent white (subspecies *magnirostris*). Confusable with *M. sapphira* (331) but which is more brilliant ultramarine on crown and rump and has a more greyish belly. Female. *Above*, olive-brown; a fulvous eye-ring. *Below*, throat and breast orange-rufous; belly buffish white. May be confused with several brown female flycatchers with rufous underparts.

Status, Habitat, etc. A rare low- and medium-elevation flycatcher. Altitudinal range little known. Affects undergrowth in shady ravines in evergreen biotope. Other details unrecorded.

340 **VERDITER FLYCATCHER** *Muscicapa thalassina* Swainson

p. 128

Size Sparrow ± ; length 15 cm (6 in.).

Field Characters Male entirely blue-green; brighter on head and throat, darker on wings and tail. A prominent black patch in front of eyes (subspecies *thalassina*). Female only duller and greener.

Status, Habitat, etc. Common summer visitor. Breeds between 1500 and 2500 m; winters in the foothills, duars and plains: light forest and wooded country, especially near streams. Singly or pairs, often among the itinerant foraging flocks. Has typical flycatcher habits. Often flutters at sprigs and flower-clusters to stampede lurking winged insects. Call: song, a pleasant jingling trill, similar to White-eye's (*Zosterops*) in pattern but louder.

341 **PIGMY BLUE FLYCATCHER** *Muscicapella hodgsoni* (Moore)

p. 129

Size Sparrow − ; length 8 cm (3 in.).

Field Characters A diminutive blue-and-orange flycatcher. Male. *Above*, dark cyan blue, brighter on crown; forehead and sides of head blue-black. *Below*, orange-yellow. Female. *Above*, olive-brown, more rufous on rump. *Below*, pale yellow (subspecies *hodgsoni*).

Status, Habitat, etc. Scarce resident, subject to vertical movements, between 2000 and 3000 m in summer, 1800 m down to the foothills in winter. Dense tall forest and secondary evergreen scrub in clearings etc. and along wooded streams. Usually solitary; lively and restless, making constant little sallies after gnats etc. within the foliage canopy or momentarily dropping to ground for creeping prey. Tail carried partly cocked between drooping wings. Food: chiefly gnat-like insects. Call: a feeble *tsip*; a distinctive high-pitched song *tzit che che che cheeee*.

342 **GREYHEADED FLYCATCHER**
Culicicapa ceylonensis (Swainson)

p. 144

Size Sparrow − ; length 9 cm (3½ in.).

Field Characters Head, neck, throat and breast ashy grey, darker on crown. Back yellowish green, rump yellow. Wings and tail brown edged with yellow. Belly bright yellow (subspecies *calochrysaea*). Sexes alike.

Status, Habitat, etc. Common summer visitor. Breeds from the foothills locally up to 3000 m; winters in the duars and widely over the Indian plains. Affects rather open but well-wooded country. Singly or pairs often among the roving mixed foraging parties. Very active and lively.

Flits incessantly and from one hunting stance to another within the foliage canopy of large trees, making agile twisting and looping sorties after winged prey and returning to its base after each capture. Food: gnats etc. Call: a very soft *pit ... pit ...* while foraging. Song, a lively high-pitched interrogative trill: *chik ... whichee-whichee?*, constantly uttered in the breeding season.

343 YELLOWBELLIED FANTAIL FLYCATCHER
Rhipidura hypoxantha Blyth   p. 129

Size Sparrow − ; length 8 cm (3 in.).

Field Characters A restless, diminutive, fantailed flycatcher. *Above*, dark greyish olive. Forehead and supercilium yellow; a broad black band through eye. Tail brown, conspicuously white-shafted and white tipped. *Below*, bright yellow. Female similar, only the eye-band blackish olive instead of black.

Status, Habitat, etc. Common altitudinal migrant. Breeds between 1800 and 3600 m; winters in the foothills, duars and adjacent plains: various types of forest and secondary jungle—preferably moist-deciduous biotope. Singly or pairs, often in the mixed foraging parties. Extremely active and lively, flitting about, making agile flycatching sallies, and pirouetting on its perch with tail characteristically fanned out and partially cocked. (The illustration is misleading in this respect.) Also flutters in front of flowers and sprigs in search of insects. Food: gnats, flies, etc. Call: a very thin, high-pitched *sip*, *sip* constantly uttered. Song, a rapid repetition of the call-note.

344 WHITETHROATED FANTAIL FLYCATCHER
Rhipidura albicollis (Vieillot)   p. 144

Size Bulbul − ; length 17 cm (6½ in.).

Field Characters A dark slaty brown fantailed flycatcher with a prominent white semi-collar across throat and short white supercilia. Outer rectrices tipped whitish (subspecies *albicollis* and *stanleyi*). Sexes alike.

Status, Habitat, etc. Common resident, subject to slight vertical movements. Duars, foothills and locally up to 2000 m or a little higher: secondary forest, groves, shrubbery, gardens, etc. Singly or pairs, often with the mixed hunting parties. Flits tirelessly amongst bushes and middle storey, pirouetting with fanned out tail and drooping wings and making sprightly looping and twisting sorties in the air after insects. Food: gnats, flies, and similar insects. Call: a harsh *chuck* or *chuck-r*. Song, feeble, jerky, of five descending whistling notes *tri-riri-riri*.

MONARCH FLYCATCHERS:
Monarchinae

345 **PARADISE FLYCATCHER** *Terpsiphone paradisi* (Linnaeus)

p. 144

Size Bulbul; length 20 cm (8 in.): male with tail ribbons up to 50 cm long.

Field Characters Male. Entire crested head and throat glistening blue-black. Bill and eye-rim blue. Wings black and white. Rest of plumage including tail-ribbons silvery white, finely black-streaked above and with black shafts to rectrices. Female and young male: crown blue-black, crest shorter. Rest of upperparts rufous with an olive wash. Nuchal collar, sides of head and throat ashy grey fading into white belly. No tail ribbons (subspecies *saturatior*).

Status, Habitat, etc. Resident at low elevations, subject to local movements. Duars and foothills up to 800 m: thin forest, secondary growth, bamboo jungle, shady gardens, etc. Typical flycatcher. Makes nimble looping sallies after winged insects. Flight swift and undulating, the long tail-ribbons rippling gracefully behind. Food: insects. Call: a nasal grating *che* or *chēchwē*. Song, a low pleasant warble of six to eight descending bulbul-like notes.

346 **BLACKNAPED MONARCH FLYCATCHER**
Monarcha azurea (Boddaert)   p. 144

Size Sparrow; length 16 cm (6 in.).

Field Characters Male. Partially fan-tailed, azure blue with whitish belly. A velvety black patch on nape and a thin black semi-gorget on throat. Female. *Above*, crown blue; rest of upperparts brown. *Below*, bluish ashy fading to whitish on belly. No nape-patch or gorget (subspecies *styani*).

Status, Habitat, etc. Resident, subject to local and winter-summer movements. Duars and foothills, locally up to 1000 m: mixed deciduous or secondary evergreen forest, and bamboo jungle. Well-wooded country, tea gardens, etc. Singly or pairs often associated with other flycatchers etc. in the mixed hunting parties. Active and restless. Pivots on its perch from side to side, prances and flits about with wings drooping and tail partly fanned and cocked. Food: winged insects. Call: a distinctive high-pitched rasping *sweech-which?* or *sweech-which-which?* (ending interrogatively). No song as such.

WARBLERS: Sylviinae

347 YELLOWBROWED GROUND WARBLER
Tesia cyaniventer Hodgson   p. 144

Size Sparrow – ; length 9 cm (3½ in.).

Field Characters A tiny dark-coloured almost tailless ground-living bird. *Above*, olive-green; a long yellowish supercilium. A distinct black stripe behind eye between the supercilium and grey sides of head. *Below*, ashy grey. Sexes alike.

Status, Habitat, etc. Resident, subject to vertical movements. Breeds between 1500 and 2500 m; winters from 1800 m down through the duars: dank shady ravines with dense undergrowth of ferns and nettles in moist-deciduous and evergreen forest. Keeps singly, skulking amongst the rootstocks, rarely showing itself. Extremely restless and inquisitive. Food: small insects, spiders. Call: a characteristic sharp *tsik* or *tchirik*. Song, a clear rippling whistle *pip-pipy-pip, pippety-pip*.

348 SLATYBELLIED GROUND WARBLER
Tesia olivea (McClelland)

Size Sparrow – ; length 9 cm (3½ in.).

Field Characters A tiny dark wren-like bird, very similar to 347. *Above*, crown yellowish green; an indistinct dark post-ocular stripe; back dark olive-green. *Below*, dark slaty. Sexes alike.

Status, Habitat, etc. Resident, subject to vertical movements. Summer altitudinal range imperfectly known; winters from at least 1000 m down through the duars: dense undergrowth of ferns, nettles and weeds in humid tropical forest. Habits, behaviour and food as in 347. Call: a distinctive sharp *tchirik-tchirik*.

349 CHESTNUTHEADED GROUND WARBLER
Tesia castaneocoronata (Burton)   p. 96

Size Sparrow – ; length 8 cm (3 in.).

Field Characters Tiny and stub-tailed, like 347 and 348. *Above*, forehead, crown and nape bright chestnut; rest olive-green. *Below*, throat bright lemon-yellow; breast and belly olive-yellow; flanks olive. Sexes alike.

Status, Habitat, etc. Resident, subject to vertical movements. Breeds between 1800 and 3300 m; winters from 1800 m down through the duars: dense undergrowth of ferns, nettles, etc. in humid tropical forest. Habits, behaviour and food as in 347, 348. Call: a chattering *tchirik-tchirik* like 347. Song, of 4 loud, shrill notes, not unlike that of Greyheaded Flycatcher (*Culicicapa ceylonensis*); reminiscent also of Yellowbrowed Ground Warbler's.

350 **PALEFOOTED BUSH WARBLER**
    Cettia pallidipes (Blanford)   p. 145
Size Sparrow − ; length 10 cm (4 in.).
Field Characters Superficially like a leaf warbler (*Phylloscopus*). *Above*,
brown; a pale supercilium and noticeable dark stripe through eye.
Below, cream-coloured, with pale yellowish legs (subspecies *pallidipes*).
Sexes alike. Like similar nondescript bush warblers most easily
recognized by its distinctive song.
Status, Habitat, etc. Resident, subject to vertical movements. Believed
to breed between 1200 and 1500 m; winters in the lower foothills, duars
and adjacent plains: grass and bush jungle on the edge of forest. Keeps
singly to low scrub. An adept skulker, often heard rarely seen. Easily
overlooked when silent. Food: insects. Call: a fast-repeated *paree-
choop* and *riti-jee*. Song, a loud and persistent whistle *rip ... rip-
chick-a-chuck*.

351 **STRONGFOOTED BUSH WARBLER** *Cettia fortipes* (Hodgson)
Size Sparrow − ; length 11 cm (4½ in.).
Field Characters *Above*, dark rufous olive-brown; a long narrow buff
supercilium; a brown eye-stripe. *Below*, dull whitish washed with
olive-brown on sides of breast and with fulvous on flanks. Legs brownish
(*v.* yellowish in *C. pallidipes*) (subspecies *fortipes*). Sexes alike.
Status, Habitat, etc. Altitudinal migrant. Breeds between 2000 to 3300 m;
winters from 2100 down through the duars and adjacent plains: bush
jungle, tea gardens, forest undergrowth, often on swampy ground.
Solitary; a confirmed skulker, hopping among bushes close to ground,
rarely showing itself. Recognized chiefly by its distinctive song. Food:
insects. Call: song, a thin prolonged ascending whistle *wheeeeee*
(*c.* 1 second) followed by a loud explosive *chiwiyou*, thus: *wheeeeee ...
chiwiyou*, with variations.

352 **LARGE BUSH WARBLER** *Cettia major* (Moore)
Size Sparrow − ; length 13 cm (5 in.).
Field Characters The highest-altitude bush warbler: dark olive-brown with
chestnut forehead and crown. A long whitish supercilium, rusty
anteriorly. Underparts dull whitish (subspecies *major*). Sexes alike.
Status, Habitat, etc. Resident, apparently subject to vertical movements.
Found in summer between 3300 and 4000 m; winter range imperfectly
known: dense rhododendron growth near tree-line, dwarf thickets
beyond. Solitary. A great skulker like other bush warblers. Food:
insects. Call: unrecorded.

353 ABERRANT BUSH WARBLER *Cettia flavolivacea* (Hodgson)

p. 144

Size Sparrow — ; length 13 cm (5 in.).

Field Characters Superficially like a leaf warbler (*Phylloscopus*) with yellowish underparts and no wing-bars. *Above*, olive-brown; a pale supercilium and dark eye-stripe. Sides of head mottled yellow and brown. *Below*, dull fulvous yellow, washed with olive on throat and flanks (subspecies *flavolivacea*). Sexes alike.

Status, Habitat, etc. Common resident, subject to vertical movements. Breeds between 2400 and 3600 m; winters from 2700 down to at least 700 m: long grass, thick scrub, ferns and undergrowth in forest. Solitary skulker, difficult to see. **Food**: insects. **Call**: song, a short, very high and thin whistle.

354 HUME'S BUSH WARBLER *Cettia acanthizoides* (Verreaux)

Size Sparrow — ; length 11 cm (4½ in.).

Field Characters A skulking rufous-brown warbler with pale yellow (greyish buff) underparts. Broad whitish supercilium and dark stripe through eye. Wings and tail brighter rufous than rest of upperparts (subspecies *brunnescens*). Sexes alike.

Status, Habitat, etc. Resident, local, subject to vertical movements— between 2400 and 3300 m; recorded elsewhere (Garhwal) down to 1350 m in winter. In dense stands of ringal bamboo (*Arundinaria*). Solitary, secretive, seldom showing itself. **Food**: insects. **Call**: song, very distinctive—a series of 3 or 4 long-drawn thin ascending whistles (each *c*. 2 seconds), uttered slowly and deliberately, followed by several quick-repeated *chew chew* up-and-down notes.

355 RUFOUSCAPPED BUSH WARBLER
Cettia brunnifrons (Hodgson)

p. 144

Size Sparrow — ; length 10 cm (4 in.).

Field Characters *Above*, rufous brown with chestnut crown, long buff supercilium, and dark eye-stripe. *Below*, throat and belly white; breast and sides grey; flanks, and vent olive-brown (subspecies *brunnifrons* and *muroides*). Sexes alike.

Status, Habitat, etc. Common altitudinal migrant. Breeds between 2700 and 4000 m; winters from 2200 down through the foothills: dwarf rhododendron, *Berberis* bushes and brackens in summer; forest undergrowth, tea gardens and shaded grassy areas in winter. Characteristic skulking habits. **Food**: insects. **Call**: alarm-note, a shrill piercing whistle. Song, a peculiar loud continuous *sip-ti-ti-sip* followed by a sound like someone blowing through a comb.

356 **SPOTTED BUSH WARBLER** *Bradypterus thoracicus* (Blyth) p. 96
Size Sparrow − ; length 13 cm (5 in.).
Field Characters A plain brown skulking warbler. *Above*, dark rufous brown; supercilium and sides of head grey. *Below*, chin white; throat ashy spotted with blackish brown; breast and belly white; flanks olive-brown. Under tail-coverts white-barred olive-brown (subspecies *thoracicus*). Sexes alike.
Status, Habitat, etc. Fairly common altitudinal migrant. Summers between 3300 and 4300 m; winters down to the duars and adjacent plains: dwarf juniper and rhododendron shrubbery, rank grass and bushes near timber-line in summer; heavy grass jungle and reed-beds in winter. A skulker like other bush warblers. Food: insects. Call: song, a persistent *see-see* note uttered as the bird rises a short distance in the air and drops again into cover.

357 **BROWN BUSH WARBLER** *Bradypterus luteoventris* (Hodgson)
Size Sparrow − ; length 13 cm (5 in.).
Field Characters *Above*, rufous-brown; a short pale supercilium and pale eye-ring. *Below*, chin, throat and belly white tinged with buff on sides, the throat sometimes with fine dark specks. Upper breast, flanks and vent rufous-brown (subspecies *luteoventris*). Sexes alike.
Status, Habitat, etc. Altitudinal migrant, locally common. Breeds between 2100 and 3300 m; winters down in the foothills: high grass-and-bracken covered hillsides. Solitary and skulking, as other bush warblers. Food: insects. Call: song, a peculiar grating trill, deceptively like a grasshopper's: of two rapidly repeated notes, a screechy *creee* followed by a short sharp *ut*, thus: *creee-ut, creee-ut* . . . .

358 **YELLOWHEADED FANTAIL WARBLER**
    Cisticola exilis (Vigors & Horsfield)
Size Sparrow − ; length 10 cm (4 in.).
Field Characters A diminutive warbler of grassy hillslopes. Male. (summer) *Above*, crown unstreaked pale orange-yellow; back rufous-brown boldly black-streaked. Tail black, tipped with buff. *Below*, centre of belly white; rest ochraceous buff. Female and winter male. *Above*, crown and back rufous streaked with black. Tail brown above, greyish below, tipped with buff and subtipped black (subspecies *tytleri*).
Status, Habitat, etc. Resident, locally common. Breeds from base of hills up to 1000 m: tall grass areas. Loose parties usually hidden in cover, individuals mounting a grass stem from time to time when tail constantly flicked open like a fan. Displaying males fly about in erratic undulating zigzags a few metres up, and nose-dive into cover. Food:

Plate 30, artist Robert Scholz

REDSTART, GRANDALA, ROBIN, SHORTWING, THRUSHES, COCHOAS

1 R **BLUEHEADED REDSTART,** *Phoenicurus caeruleocephalus*   page 196
Sparrow ±. ♀ Brown. White wing-patch. Rump rufous. Tail brown, edged rufous. Belly and vent white. 1200–3900 m.

2 R **GRANDALA,** *Grandala coelicolor*   201
Bulbul +. ♀ Brown, streaked whitish. Rump tinged blue. A white wing-patch. 3000–5400 m.

3 R **BLUEFRONTED ROBIN,** *Cinclidium frontale*   201
Bulbul ±. ♂♀ Like 415 but tail graduated, without basal white patches. Recorded at *c.* 2250 m.

4 R **GOULD'S SHORTWING,** *Brachypteryx stellata*   190
Sparrow −. 2000–4200 m.

5 R **CHESTNUTBELLIED ROCK THRUSH,** *Monticola rufiventris*   206
Myna ±. ♀ Like 430 but with whitish throat-patch. Duars to 3300 m.

6 R **SMALLBILLED MOUNTAIN THRUSH,** *Zoothera dauma*   210
Myna +. Large buff underwing-patch in flight. Duars to 3400 m.

7 R **PURPLE COCHOA,** *Cochoa purpurea*   203
Myna +. 1000–3000 m.

8 R **GREEN COCHOA,** *Cochoa viridis*   203
Myna +. 700–1500 m.

Inches

0   4   8

R. Scholz

Plate 31, artist Winston Creado

FORKTAILS, REDSTART, THRUSHES, BUSH CHAT

1 R   LITTLE FORKTAIL, *Enicurus scouleri*   page 201
Sparrow — . 300–3300 m.

2 R   SLATYBACKED FORKTAIL, *Enicurus schistaceus*   202
Bulbul. Up to 1600 m.

3 R   SPOTTED FORKTAIL, *Enicurus maculatus*   203
Bulbul. Up to 3000 m.

4 R   BLACKBACKED FORKTAIL, *Enicurus immaculatus*   202
Bulbul. Up to 1450 m.

5 R   WHITECAPPED REDSTART, *Chaimarrornis
leucocephalus*   205
Bulbul. Up to 5100 m.

6 R   WHISTLING THRUSH, *Myiophonus caeruleus*   206
Pigeon. Yellow bill. Black legs. Up to 4000 m.

7 R   DARK-GREY BUSH CHAT, *Saxicola ferrea*   204
Sparrow. ♀ Rusty rump. Rufous-edged tail. Up to
3300 m.

8 R   ORANGEHEADED GROUND THRUSH, *Zoothera
citrina*   207
Myna — . White wing-bar. ♀ Tinged olive-brown
above. Up to 1600 m.

9 R   BLUE ROCK THRUSH, *Monticola solitarius*   206
Bulbul + . ♀ Back feathers with dark shaft-streaks.
Up to 3000 m.

10 R BLUEHEADED ROCK THRUSH, *Monticola
cinclorhynchus*   205
Bulbul. ♀ Back plain olive brown. Up to 2200 m.

insects. Call: song given in display, a wheezy *scrrrrrr* followed by an explosive liquid, bell-like *plook*, thus: *scrrrrrr ... plook* repeated *ad lib.*

359 RUFOUS WREN-WARBLER *Prinia rufescens* Blyth   p. 144

Size Sparrow − ; length 11 cm (4½ in.).

Field Characters A slim, skulking brown warbler with long, graduated black-and-white-tipped tail, carried half-cocked and loosely switched up and down. *Above*, rufous-brown; a buff supercilium and pale eye-ring. Crown and nape ashy brown in summer, nearly concolorous with back in winter. *Below*, pale buff; more ochraceous on flanks and belly (subspecies *rufescens*). Sexes alike.

Status, Habitat, etc. Common resident. Breeds in the foothills and normally up to 1200 m, rarely to 2000 m: long grassland, low scrub and weeds, and edges of secondary jungle. Singly, pairs or small parties in low herbage, seldom showing themselves. Takes short jerky flights and dives into cover. Food: insects. Call: a feeble monotonous *seep* or *chip-wee* constantly repeated. Song, rather similar to 360, q.v.

360 ASHY-GREY WREN-WARBLER *Prinia hodgsonii* Blyth   p. 144

Size Sparrow − ; length 11 cm (4½ in.).

Field Characters A small rufous-brown, long-tailed warbler very similar to *P. rufescens* (359); distinguished from it by the grey (*v.* rufous) tail and, in summer, by a cloudy grey band across the breast (subspecies *rufula*). Sexes alike.

Status, Habitat, etc. Common resident, locally subject to downward movement in winter: plains, duars and foothills, normally up to 1200 m—scrub-and-grass jungle, shrubbery around cultivation and villages, etc. Singly, pairs, or parties sometimes of 15 or 20. Works through undergrowth and foliage of small trees. Flits jerkily from bush to bush, constantly flicking tail, the birds loosely following one another. Food: insects. Call: an incessant conversational tinkling *zee-zee-zee*. Song, a vehement squeaky *yousee-yousee-yousee-which-which-which-which* starting feebly, rising in loudness and tempo and ending abruptly; repeated several times.

361 HODGSON'S WREN-WARBLER *Prinia cinereocapilla* Hodgson

Size Sparrow − ; length 11 cm (4½ in.).

Field Characters *Above*, forehead and narrow supercilium rufous (sometimes absent); crown, nape, and sides of neck dark grey; rest dark rufous. Tail rufous-brown with a subterminal blackish patch on rectrices. *Below*, fulvous, darker on lower flanks and vent. Sexes alike. Dark grey cap and very rufous back and wings diagnostic.

Status, Habitat, etc. Fairly common but little known; from the edge of the plains through the duars and foothills up to at least 1350 m: dense jungle and secondary growth. Habits as of other wren-warblers. Food: insects, flower nectar. Call: song, a rising trill ending in a long-drawn *swe-e-e-e-e-chor*, like an Iora's; better and more varied than other *Prinias*'.

362 PLAIN WREN-WARBLER *Prinia subflava* (Gmelin)   p. 144

Size Sparrow − ; length 13 cm (5 in.).

Field Characters *Above*, sandy brown. A narrow supercilium, sides of head, and outer rectrices whitish. *Below*, cream-coloured—more yellowish in fresh nuptial plumage (subspecies *fusca*). Sexes alike.

Status, Habitat, etc. Common resident, from the edge of the plains up to 1200 m in the hills: high grass in open fields, secondary growth and bamboo-and-scrub jungle. Pairs or small parties. Habits and food typical of wren-warblers. In the peculiar jerky switchback flight the longish loosely swinging tail looks as though too heavy for the bird. Call: a rather plaintive *tee-tee-tee*. Song, wheezy, grasshopper-like, rapid *tlick-tlick-tlick* ... sustained for ten seconds or more and constantly repeated.

363 ASHY WREN-WARBLER *Prinia socialis* Sykes

Size Sparrow − ; length 13 cm (5 in.).

Field Characters *Above*, head, sides of neck and back slaty grey: rest of upperparts warm rufous-brown. Tail graduated, the rectrices with fulvous tips and blackish subterminal spots. *Below*, throat buffy white; rest of underparts strongly washed with rufous, especially on flanks and lower belly (subspecies *inglisi*). Sexes alike.

Status, Habitat, etc. Common resident. Plains, duars and foothills up to 1200 m: grassland, secondary jungle, cultivation, gardens, etc. Habits and food as of other wren-warblers. Call: a sharp, rather nasal *tee-tee-tee*. Song rather wheezy and chipping—a loud sprightly *jimmy-jimmy-jimmy* quickly repeated five or six times.

364 BROWN HILL WARBLER *Prinia criniger* Hodgson   p. 144

Size Sparrow − , with a long tail; length 16 cm (6 in.).

Field Characters A long-tailed brown warbler with dark streaked upperparts—a passable miniature of the Common Babbler (*Turdoides caudatus*). Tail steeply graduated, the rectrices buff-tipped with a dusky subterminal spot. *Below*, pale fulvous (summer) or mottled with dusky on sides of throat and breast (winter); flanks olive-brown (subspecies *criniger*). Sexes alike.

Status, Habitat, etc. Common resident, subject to small vertical move-

ments. Breeds between 1200 and 2800 m (locally); winters down to
300 m in the foothills: bush-and-grass-covered hillsides, edge of terraced
cultivation and open pine forest. Singly or pairs, skulking in grass
tussocks and low scrub. Flight feeble and jerky, as of wren-warblers.
Food: insects. Call: song, a series of several lively but wheezy double
notes like a knife being sharpened on a grindstone: *tsee-tswee-tsee-
tswee* ... Given from a perch or during the 'dive-bombing' aerial
display.

365 BLACKTHROATED HILL WARBLER

Prinia atrogularis (Moore)   p. 144

Size Sparrow − , with a long tail; length 17 cm (6½ in.).

Field Characters Summer. *Above*, unstreaked dark olive-brown, greyer on
head and nape. Wings and tail rufous-brown. Tail steeply graduated,
up to 10 cm long, paler at tip. *Below*, throat black bordered by white
moustachial stripes; breast spotted black and white; belly fulvous,
flanks more olive. In winter throat fulvous-white streaked with black;
a distinct white supercilium (subspecies *atrogularis*). Sexes alike.

Status, Habitat, etc. Fairly common resident, but local. Breeds between
1000 and 2500 m; winters at somewhat lower elevations. Habitat as of
364 and wren-warblers. Call: alarm, a series of soft, scolding *churrrr-
churrrr-churrrr*. Song, loud, rather similar to that of 364, also rendered as
tulip-tulip-tulip ...

366 TAILOR BIRD *Orthotomus sutorius* (Pennant)   p. 144

Size Sparrow − ; length 13 cm (5 in.).

Field Characters A small restless yellowish green warbler with rust-
coloured crown and whitish buff underparts. Tail graduated, with the
central rectrices long and narrow, normally carried jauntily erect
(subspecies *patia* and *luteus*). Sexes alike: tail usually longer in male.

Status, Habitat, etc. Common resident, from the edge of the plains
through the duars and foothills up to 1800 m: herbaceous gardens,
hedgerows, orchards, secondary jungle, etc. Singly or pairs, hopping
about with cocked tail, in bushes or on the ground, uttering its loud
familiar calls. Food: insects, flower nectar. Call: alarm, a loud
quick-repeated *pit-pit-pit-pit* ... Song, a surprisingly loud *pitchik-
pitchik-pitchik* ... or *chubit-chubit-chubit* ... given in long runs.

367 GOLDENHEADED TAILOR BIRD

Orthotomus cucullatus Temminck   p. 145

Size Sparrow − ; length 12 cm (5 in.).

Field Characters *Above*, forehead and crown rufous; a short yellow

supercilium. Sides of head and a nuchal collar dark grey; back and
wings olive-green; rump yellow. Tail brown, outermost rectrices
partly white. *Below*, throat and breast pale grey; rest bright yellow
(subspecies *coronatus*). Sexes alike.

Status, Habitat, etc. Resident, locally common, from the duars up to
1800 m: evergreen biotope—climax forest, secondary growth, bamboo
jungle and scrub. Habits etc. very similar to Tailor Bird (366) but is
exclusively forest-dwelling. Hunts in low thickets; elusive and difficult
to observe. Food: insects. Call: song, a loud 4-noted whistle *pee-pi-pi-
pee* repeated in a number of different keys.

368 STRIATED MARSH WARBLER *Megalurus palustris* Horsfield
Size Bulbul + ; length 25 cm (10 in.).

Field Characters A very large black-streaked warbler with long, pointed,
graduated tail. *Above*, fulvous brown with fine streaks on head, broad
black stripes on back and wings; a pale supercilium. Tail fulvous-brown.
Below, fulvous-white (dull primrose yellow in fresh plumage), finely
brown-streaked on breast, flanks and vent (subspecies *toklao*). Sexes
alike.

Status, Habitat, etc. Resident, locally common. Duars and lower
foothills: reed-beds on swampy ground, tall grass and scrub in
abandoned clearings and pond environs. Singly or pairs. Less skulking
than other grass warblers; mounts to exposed bush-tops early mornings.
When perched upright looks like a shrike; flicks tail loosely up and
down like a chat. Sometimes descends to ground, walking about to
feed (not hopping). Food: insects. Call: a clear, long-drawn subdued
whistle ending in a short explosive *wheeechoo*, of pattern of a Bush
Warbler's (*Cettia*) call. Also other notes, and a Calandra Lark-like
song.

369 BLYTH'S REED WARBLER *Acrocephalus dumetorum* Blyth
Size Sparrow − ; length 14 cm (5½ in.).

Field Characters *Above*, olive-brown; a pale supercilium. *Below*, throat
white, rest buffish. Sexes alike.

Status, Habitat, etc. Common winter visitor. Lower foothills, duars and
adjacent plains: deciduous lightly wooded country—shrubbery, hedges,
gardens, orchards, bamboo thickets, etc. Keeps singly, hopping and
creeping through undergrowth unseen, its presence betrayed only by
its loud call-note. Food: insects. Call: a harsh single *chuck*, like
clicking one's tongue against the palate, uttered every few seconds
while foraging. Song, a subdued melodious warbling *chek-chek-chek-*

che-chwee-chek-chek rapidly repeated, punctuated with sundry harsh *chuck*s and *chur-r*s.

370 **BROWN CHIFFCHAFF** *Phylloscopus collybita* (Vieillot)

Size Sparrow − ; length 10 cm (4 in.).

Field Characters No wing-bar. *Above*, pale olive-brown, a short whitish supercilium. *Below*, dull whitish, washed with buff on breast and flanks. Legs and feet dark brown (subspecies *tristis*). Sexes alike.

Status, Habitat, etc. Abundant winter visitor, from 2100 m down through the foothills, duars and adjacent plains: bushes, hedges, groves, scrub jungle, standing crops, etc. Singly or parties of 8 to 10 or more. Flits restlessly from bush to bush, or on ground, constantly flicking wings and tail as characteristic of the genus. Food: insects. Call: a plaintive *tweet* or *chivit*. Song, a rapid repetition of the call-note *chi-vit chi-vit* connected by a few musical notes.

371 **TICKELL'S LEAF WARBLER** *Phylloscopus affinis* (Tickell)

p. 160

Size Sparrow − ; length 10 cm (4 in.).

Field Characters No wing-bar. *Above*, dark olive-brown; a prominent long yellow supercilium. *Below*, almost canary yellow (subspecies *affinis*). Sexes alike.

Status, Habitat, etc. Abundant summer visitor. Breeds at high altitudes, between 2700 and 4800 m; winters down through the foothills, duars and adjacent plains. *Berberis*, juniper, rhododendron bushes (summer); well-wooded country, secondary jungle (winter). Pairs in summer; loose parties in winter often among the mixed foraging bands. Usually feeds in low bushes, flicking wings and tail in the characteristic manner. Food: insects. Call: song, a single note rapidly repeated 5 or 6 times preceded by one or two higher-pitched ones, thus *pick-pick whi whi whi whi whi*.

372 **SMOKY LEAF WARBLER** *Phylloscopus fuligiventer* (Hodgson)

p. 160

Size Sparrow − ; length 10 cm (4 in.).

Field Characters No wing-bar. *Above*, very dark olive-brown; a long greenish yellow supercilium. *Below*, dull greenish yellow or dusky oil-green (subspecies *fuligiventer* and *tibetanus*). Sexes alike.

Status, Habitat, etc. Altitudinal migrant. Breeds from 3600 m to over 4300; winters down through the foothills, duars and adjacent plains. Low scrub above timber-line and boulder-strewn alpine meadows (summer); dense bushes, grass and sedges near stream etc. (winter).

Keeps in low scrub or on stony ground, clambering among boulders for food; in winter usually near water. Typical leaf warbler in habits and food. Call: song, a single or double note rapidly and monotonously repeated 5 or 6 times, thus: *tsli* (or *tsuli*)-*tsli-tsli-tsli* ...

373 DUSKY LEAF WARBLER *Phylloscopus fuscatus* (Blyth)   p. 96

Size Sparrow − ; length 10 cm (4 in.).

Field Characters No wing-bar. *Above*, dusky olive-brown; a prominent greyish supercilium and semi eye-ring, and a dark streak through eye. *Below*, greyish white; flanks and vent fulvous (subspecies *weigoldi*). Sexes alike.

Status, Habitat, etc. Winter visitor. Foothills, duars and adjacent plains: reed and scrub jungle, low bushes and long grass around pools, and crops in damp areas. Usually forages singly in low bushes or on the ground, hopping about, incessantly flicking wings and tail. Food: insects. Call: a sharp clicking note *chuck*, like Lesser Whitethroat's but softer.

374 ORANGEBARRED LEAF WARBLER *Phylloscopus pulcher* Blyth
p. 96

Size Sparrow − ; length 10 cm (4 in.).

Field Characters Two orange-yellow wing-bars, the lower very prominent, thus one of the most easily identified leaf warblers. *Above*, olive; crown sooty olive with an indistinct median stripe and prominent yellow supercilium to nape. Rump pale yellow; outer rectrices partly white. *Below*, pale yellow, greyish on breast (subspecies *pulcher*). Sexes alike.

Status, Habitat, etc. Altitudinal migrant. Breeds between 3000 and 4300 m; winters lower, between 2800 and 500 m: mixed conifer and birch forest, and dwarf rhododendron scrub above tree-line (summer); oak and other low-elevation forest (winter). More arboreal than many other leaf warblers; in winter commonly among the mixed hunting parties of tinies high up in canopy foliage. Food: insects. Call: a thin, high-pitched *tsip* constantly repeated, sometimes quickly run together like a flowerpecker's twitter.

375 YELLOWBROWED LEAF WARBLER
Phylloscopus inornatus (Blyth)   p. 160

Size Sparrow − ; length 10 cm (4 in.).

Field Characters Two buffish wing-bars, upper less distinct; tail relatively short. *Above*, greenish olive; faint coronal bands (two dark lateral, one buffish median); prominent yellowish supercilium and eye-ring. *Below*, yellowish white (subspecies *mandellii*). Sexes alike.

Status, Habitat, etc. Winter visitor; from 1800 m down through the foothills, duars and adjacent plains. May breed in Arunachal Pradesh above 2800 m—open forest and wooded country. Forages in canopy foliage of high trees as well as bushes near ground. Flutters against sprigs and flowers or makes short flycatching sallies in air. Constantly flicks tail and wings. Food: insects. Call: a sparrow-like *chilp*; also a sharp double note *tiss-yip*. Song, described as *fitifitifiti* mixed with trills.

376 **PALLAS'S LEAF WARBLER** *Phylloscopus proregulus* (Pallas)

p. 160

Size Sparrow − ; length 9 cm (3½ in.).

Field Characters Two yellow wing-bars. *Above*, olive-green, with a prominent yellow rump-band. Two dusky olive coronal bands with a prominent yellowish median stripe ('centre parting'); a long conspicuous yellow supercilium and dark streak through eye. *Below*, dull yellowish white (subspecies *newtoni*). Sexes alike. Easily identified by yellow rump when fluttering in front of sprigs to take insects.

Status, Habitat, etc. Altitudinal migrant, common. Breeds between 2700 and 4200 m; winters from 2100 m down through the foothills and duars: coniferous forest mixed with birch, rhododendron, etc. in summer; jungle-covered hillsides and wooded country in winter. Generally keeps high up in trees, often with the mixed hunting parties. Extremely active and restless. Food: insects. Call: a soft double *tsip-tsip*. Song, of various notes repeated 4 or 5 times.

377 **GREYFACED LEAF WARBLER**
   Phylloscopus maculipennis (Blyth)   p. 160

Size Sparrow − ; length 9 cm (3½ in.).

Field Characters Double yellow wing-bar, combined with small size, grey head and throat, yellow rump, and white in shortish tail identifies this species. *Above*, olive-green, rump yellow. Head brownish grey with a whitish median coronal stripe; a long prominent supercilium and a dark streak through eye; cheeks mottled grey. *Below*, throat and breast grey; rest yellow (subspecies *maculipennis*). Sexes alike.

Status, Habitat, etc. Resident, subject to vertical movements. Breeds between 2500 and 3400 m; winters from 2900 down through the foothills and duars: open conifer and mixed forest in summer; oak and other broad-leaved forest and secondary scrub in winter. Habits and food as in other leaf warblers. Call: a short *zitt*, very similar to that of *proregulus* (376). Song, undescribed.

378 LARGEBILLED LEAF WARBLER
Phylloscopus magnirostris Blyth   p. 160
Size Sparrow − ; length 12 cm (5 in.).

Field Characters A large leaf warbler with single indistinct wing-bar. *Above*, brownish olive; a prominent yellowish supercilium and dark eye-streak; cheeks mottled yellow and dusky. *Below*, yellowish white, tinged with grey on breast and throat. Sexes alike. Very similar to Greenish Leaf Warbler (379); distinguished from it by calls, q.v.

Status, Habitat, etc. Fairly common summer visitor. Breeds between 1800 and 3600 m; winters mostly in the Peninsula: conifer and mixed forest, chiefly vicinity of streams (summer); humid evergreen biotope (winter). Singly in leafy middle storey. Presence usually detected by call-note and song. Food: insects. Call: a distinctive interrogative *dir-tee?* or *weee-chi?*. Song, clear, 5-noted silver-bell-like *see-sisi-sisi* on 3 descending tones; reminiscent of ditty of Whitethroated Fantail Flycatcher (344).

379 GREENISH LEAF WARBLER
Phylloscopus trochiloides (Sundevall)   p. 96
Size Sparrow − ; length 10 cm (4 in.).

Field Characters A single faint wing-bar. *Above*, dull greenish. A well-marked yellow supercilium and dark eye-streak; cheeks mottled olive-yellow. *Below*, sullied yellowish white (subspecies *trochiloides*). Sexes alike. Distinguished from similar Largebilled (378) by calls.

Status, Habitat, etc. Summer visitor. Presumably breeds between 2700 and 3700 m as in western Himalayas; winters chiefly in the duars and the Peninsula: open deciduous jungle, wooded compounds, orchards, village groves, etc. Arboreal, foraging chiefly in middle storey. Habits and food typical. Call: a squeaky *chiwee* or *si-chiwee* (sometimes *si-si-si-chiwee*) repeated every few seconds. Song, a lively, high-pitched, quick-repeated *chi-chirichi-chiwichee* and variants.

380 CROWNED LEAF WARBLER *Phylloscopus occipitalis* (Blyth)
     p. 160
Size Sparrow − ; length 10 cm (4 in.).

Field Characters Two wing-bars, the upper indistinct. *Above*, light greyish olive, yellower on wings. Crown dusky olive with a pale median stripe; a prominent pale yellow supercilium and dark eye-streak; cheeks pale yellow. *Below*, whitish, with contrasting yellow under tail-coverts (subspecies *coronatus*). Sexes alike.

Status, Habitat, etc. Rare winter visitor: duars and foothills up to

2000 m (Sikkim). Keeps to lower branches of trees, and bushes. Very silent and unobtrusive. Call: not recorded in winter.

381 BLYTH'S CROWNED LEAF WARBLER
Phylloscopus reguloides (Blyth)   p. 160

Size Sparrow — ; length 9 cm (3½ in.).

Field Characters Very similar to 380 but smaller; under tail-coverts concolorous with whitish underparts, not yellow (subspecies *reguloides* and *assamensis*). Sexes alike.

Status, Habitat, etc. Altitudinal migrant, fairly common. Breeds between 2000 and 3500 m; winters from 1500 m down through the foothills and duars: mixed conifer and rhododendron forest in summer; bush country with scattered trees in winter. Habits, food, and general ecology typical. Call: a squeaky *kee-kew-i* constantly repeated. Song, a distinctive trill of 9 or 10 notes: *chi-ti-chi-ti-chi-ti-chi-ti-chee*.

382 YELLOWFACED LEAF WARBLER
Phylloscopus cantator (Tickell)   p. 160

Size Sparrow — ; length 9 cm (3½ in.).

Field Characters A double yellow wing-bar, the upper indistinct. *Above*, olive-green. Crown with two lateral black bands and a yellowish green median stripe; a prominent yellow supercilium; sides of head bright yellow. *Below*, throat and vent bright yellow, belly whitish (subspecies *cantator*). Sexes alike.

Status, Habitat, etc. Widespread in winter in the duars and foothills; breeding range little known. Affects dense forest in summer; more open, deciduous mixed jungle in winter. Habits, food, etc. as of other leaf warblers. Call: a loud, incessant *pio, pio* ... Song undescribed.

383 ALLIED FLYCATCHER-WARBLER   *Seicercus affinis* (Hodgson)
      p. 160

Size Sparrow — ; length 10 cm (4 in.).

Field Characters A single yellow wing-bar. *Above*, olive-green. Crown striped grey and black; sides of head yellow; a prominent whitish eye-ring. Outer tail-feathers partly white. *Below*, entirely bright yellow. Sexes alike.

Status, Habitat, etc. Resident, subject to altitudinal movements. From 2300 m (summer) down to the duars (winter): dense humid evergreen or pine forest. Habits, food, general ecology very similar to leaf warblers'. Usually present among the itinerant mixed foraging parties of tinies. Call: a short *che-wheet*. Song, not adequately recorded.

384 BLACKBROWED FLYCATCHER-WARBLER
Seicercus burkii (Burton)

p. 145

Size Sparrow — ; length 10 cm (4 in.).

Field Characters No wing-bar. *Above*, olive-green, including sides of head and supercilium; a bright yellow eye-ring; lateral coronal bands black. *Below*, deep yellow (subspecies *burkii*). Sexes alike.

Status, Habitat, etc. Common summer visitor, subject to vertical movements. Breeds between 2000 and 3700 m; winters from 2100 m down through the duars: oak, rhododendron and birch forest as well as mixed conifer and deciduous—undergrowth and middle storey. In winter often small parties among the mixed foraging assemblages of small birds. Actions and behaviour very like leaf warblers'. **Food**: insects. **Call**: a sharp single *chiw* or *wee-up* while foraging. Song, canary-like, of loud clear trilling notes.

385 GREYHEADED FLYCATCHER-WARBLER
Seicercus xanthoschistos (Gray)

p. 160

Size Sparrow — ; length 10 cm (4 in.).

Field Characters *Above*, crown and back grey; a long white supercilium; rump and wings greenish yellow; outer tail-feathers largely white. *Below*, bright yellow (subspecies *xanthoschistos* and *flavogularis*). Sexes alike. In flight tail flicked open and shut scissors-wise, flashing the white rectrices which proclaim its identity.

Status, Habitat, etc. Common resident, subject to vertical movements. Breeds between 1000 and 2700 m, occasionally lower; winters from 2000 m down through the foothills and duars: open evergreen forest, coniferous or broad-leaved. Habits and behaviour very like leaf warblers'. **Food**: chiefly insects. **Call**: a distinctive high-pitched, oft-repeated *psit-psit*, and a plaintive *tyee-tyee*. A lively song resembling that of Blackbrowed (384).

386 GREYCHEEKED FLYCATCHER-WARBLER
Seicercus poliogenys (Blyth)

p. 160

Size Sparrow — ; length 10 cm (4 in.).

Field Characters A yellow wing-bar. *Above*, olive-green; head and neck slaty with two black lateral bands on crown, and a prominent white eye-ring. Outer tail-feathers largely white. *Below*, chin and lower cheeks pale grey; rest bright yellow. Sexes alike. Greyheaded Flycatcher (342) has similar colour scheme, but its head is paler grey and without the diagnostic white eye-ring.

Status, Habitat, etc. Resident, subject to vertical movements. Altitudinal range unclear; observed between 2400 and 3000 m in spring, down to lower foothills in winter: open evergreen forest and bamboo jungle. Habits, behaviour and food like leaf warblers'. Call: unrecorded.

387 CHESTNUTHEADED FLYCATCHER-WARBLER

Seicercus castaniceps (Hodgson)   p. 145

Size Sparrow — ; length 10 cm (4 in.).

Field Characters Two yellow wing-bars. *Above*, crown chestnut with dark lateral bands becoming broader and blacker posteriorly, a white patch between them on nape; nape and sides of head grey; a prominent white eye-ring. Back and wings yellowish olive; rump bright yellow; outer tail-feathers largely white. *Below*, upper breast bluish ashy; rest bright yellow (subspecies *castaniceps*). Sexes alike.

Status, Habitat, etc. Resident, locally distributed, subject to vertical movements. Breeds between 1800 and 2400 m; winters from 2300 down to the duars: dense oak forest etc. Habits and behaviour very like leaf warblers'. Often keeps in foraging parties with them and with other small birds. Food: insects. Call: a double note *chi-chi*, and a distinctive loud wren-like *tsik*. Song undescribed.

388 YELLOWBELLIED FLYCATCHER-WARBLER

Abroscopus superciliaris (Blyth)   p. 160

Size Sparrow — ; length 9 cm (3½ in.).

Field Characters *Above*, forecrown brownish grey; a broad white supercilium; a blackish streak through eye; sides of head mottled grey. Rest of upperparts yellowish olive. *Below*, throat and upper breast whitish; rest bright yellow (subspecies *flaviventris* and *drasticus*). Sexes alike.

Status, Habitat, etc. Resident, locally common, subject to vertical movements. Breeds between 500 and 900 m; winters down to the adjacent plains: mixed bamboo forest and scrub jungle, preferably near streams. Habits etc. typically leaf warbler-like. Food: insects. Call: not intelligibly recorded.

389 BLACKFACED FLYCATCHER-WARBLER

Abroscopus schisticeps (Gray)   p. 145

Size Sparrow — ; length 9 cm (3½ in.).

Field Characters *Above*, crown, nape and ear-coverts slaty grey; prominent yellow supercilia meeting on forehead; a broad black band through eye. Rest of upperparts olive; outer tail-feathers largely white. *Below*, yellow, washed with olive on breast, paling to white on belly (subspecies *schisticeps* and *flavimentalis*). Sexes alike.

Status, Habitat, etc. Resident, locally common between 1500 and 2500 m, possibly with some vertical movements: oak and bamboo forest—in scrub-covered ravines etc. Habits like leaf warblers'. Usually met among the mixed foraging parties of small arboreal birds. Food: insects. Call: not intelligibly recorded.

390 WHITETHROATED FLYCATCHER-WARBLER
Abroscopus albogularis (Horsfield & Moore)
Size Sparrow − ; length 8 cm (3 in.).
Field Characters *Above*, head rufous with two black superciliary bands to nape. Back and wings olive-green; rump yellowish white. No white in tail-feathers. *Below*, throat and belly white separated by a yellow breast-band; vent yellow (subspecies *albogularis*). Sexes alike.
Status, Habitat, etc. Resident, locally common between 300 and 1200 m: bamboo jungle, secondary scrub, and moist-deciduous forest. Habits etc. like leaf warblers'. Usually among mixed foraging flocks of tinies. Food: insects. Call: a shrill twitter.

391 BROADBILLED FLYCATCHER-WARBLER
Abroscopus hodgsoni (Moore) p. 160
Size Sparrow − ; length 10 cm (4 in.).
Field Characters *Above*, forehead and crown chestnut; sides of head grey; a short ashy supercilium. Rest of upperparts olive-green, tinged with yellow on rump. *Below*, chin and breast dark grey, paler grey on throat. Rest bright yellow (subspecies *hodgsoni*). Sexes alike.
Status, Habitat, etc. Resident, possibly with some vertical movements. Breeds between 1100 and 2700 m—dense scrub and bamboo jungle along forest edges etc. Keeps to thick undergrowth, frequently among the mixed foraging parties. Habits and food as of leaf warblers. Call: a single long-drawn whistle followed a second or so later by two notes, the second lower.

392 GOLDCREST *Regulus regulus* (Linnaeus) p. 96
Size Sparrow − ; length 8 cm (3 in.).
Field Characters A diminutive phylloscopus-like bird with two yellowish white wing-bars. **Male.** *Above*, crown with two broad black lateral bands and an orange median stripe; a large white area round eyes. Back and wings greyish olive-green, tinged with yellowish on rump; a dark patch on wing. *Below*, pale greyish fulvous, washed with olive on flanks. **Female** similar, but with the median coronal stripe yellow (subspecies *sikkimensis* and *yunnanensis*).
Status, Habitat, etc. Resident, subject to vertical movements. From 2200 to 3500 m, locally up to 4000 m: temperate conifer forest in summer; oak and birch forest in winter. Keeps to the canopy of conifers,

often among the mixed hunting flocks, restlessly flitting or hovering among the sprigs. Behaviour and actions very like leaf warblers'. Food: insects. Call: a distinctive high-pitched mouse-like squeaking while foraging in the fir-tops.

THRUSHES, CHATS: Turdinae

393 GOULD'S SHORTWING *Brachypteryx stellata* Gould   p. 176
Size Sparrow − ; length 13 cm (5 in.).
Field Characters A short-tailed robin, with bright chestnut upperparts. *Below*, slaty grey, vermiculated and flecked with black and white. Sexes alike.
Status, Habitat, etc.   Resident, very local, subject to vertical movements. Between 3300 and 4200 m in summer; down to 2000 m or lower in winter: dense dwarf rhododendron and bamboo growth on hillsides and dank ravines with moss and fern undergrowth. Largely terrestrial, running (not hopping) in and out of roots and fallen branches like a mouse. Food: insects, seeds. Call: alarm, *tik-tik*. Song unknown.

394 RUSTYBELLIED SHORTWING
    Brachypteryx hyperythra Jerdon & Blyth   p. 161
Size Sparrow − ; length 13 cm (5 in.).
Field Characters Male. *Above*, lores and a frontal line black; a short, partly concealed white supercilium. Rest of upperparts deep blue. *Below*, entirely ferruginous. Female. *Above*, olive-brown. *Below*, pale ferruginous; centre of belly whitish. Male distinguished from similar Blue Chat (399) by less prominent white supercilium.
Status, Habitat, etc.   Endemic, rare and little known. Has been recorded in January at 1100 and 2900 m. Frequents dense *Arundinaria* bamboo and thickets. Food: unrecorded. Call: unknown.

395 LESSER SHORTWING *Brachypteryx leucophrys* (Temminck)
    p. 161
Size Sparrow − ; length 13 cm (5 in.).
Field Characters Male. *Above*, dark slaty blue. A white supercilium, often concealed. *Below*, throat and belly white; a breast-band and flanks smoky grey. Female. *Above*, rusty olive-brown. A white, usually concealed supercilium. *Below*, throat and belly white; a breast-band and flanks rusty brown (subspecies *nipalensis*).
Status, Habitat, etc.   Resident, uncommon, subject to vertical movements. Between 1500 and 3900 m in summer, down to the foothills in

winter: dense undergrowth in humid forest, preferably near streams. Singly or pairs; mainly terrestrial. Forages on the ground unobtrusively, running about among decaying tree-trunks etc. **Food:** chiefly insects. **Call:** alarm, a loud single piping note. Song, a short warble of a few notes rather like that of Whitebrowed Shortwing (396) q.v.

396 WHITEBROWED SHORTWING

Brachypteryx montana Horsfield   p. 161
Size Sparrow − ; length 13 cm (5 in.).

Field Characters Male. Entirely dark slaty blue, ashy on belly; lores velvety black; a long white supercilium. Female. *Above,* dark olive-brown; lores and eye-ring rusty. *Below,* paler olive-brown, fulvous on belly, rufescent on vent (subspecies *cruralis*).

Status, Habitat, etc.   Resident, fairly common, subject to vertical movements. Breeds between 1500 and 3300 m; recorded in winter between 300 and 2400 m: undergrowth of bracken etc. in damp, shady oak and rhododendron forest, preferably near streams. Mainly terrestrial. Solitary. Behaviour typically robin-like. **Food:** chiefly insects. **Call:** alarm-note, *tt-tt-tt* accompanied by a flick of wings. Song, 'a gentle merry little warble which may be syllabified as *Hey did-dle did-dle the cat an'*.

397 RUBYTHROAT *Erithacus calliope* (Pallas)   p. 64

Size Sparrow; length 15 cm (6 in.).

Field Characters Male. *Above,* olive-brown; no white in tail. Lores black; a conspicuous white supercilium. *Below,* chin and throat scarlet, bordered with a black line on sides of chin; a broad white malar stripe. Breast and flanks pale buffish brown; belly whitish. Female like male, also with white supercilium, but throat white or pinkish without the black line. Belly and vent buff.

Status, Habitat, etc.   Winter visitor, uncommon. Duars and foothills up to 1500 m, in dense scrub, long grass, reeds, tea gardens etc. Usually single, feeding on ground. Constantly flicks cocked and expanded tail over back. Carriage and behaviour—dipping forepart and running about in short rapid spurts—typically robin-like. **Food:** insects. **Call:** a harsh *ché* reminiscent of Jungle Babbler's, sometimes followed by pleasing snatches of a plaintive whistling song.

398 EASTERN RUBYTHROAT *Erithacus pectoralis* (Gould)

Size Sparrow ± ; length 15 cm (6 in.).

Field Characters Male. *Above,* brownish slaty; forehead and supercilium white; wings brown. Tail blackish brown with white base and tip.

192

Plate 32, artist Winston Creado

ACCENTORS, TITS, DIPPERS, THRUSHES, BLACKBIRD

Winston Creado

Inches
0   2   4

P. BARRUEL

Plate 33, artist Paul Barruel

ACCENTOR, GROSBEAK, FINCHES, SISKIN

1 R **RUFOUSBREASTED ACCENTOR,** *Prunella strophiata* page 216
Sparrow. Prominent supercilium. Black-streaked white throat. 1300–4300 m.

2 R **ALLIED GROSBEAK,** *Mycerobas affinis* 239
Myna. No wing-patch. Thigh feathers yellow. ♀ Like 507 but no wing-patch; throat grey. 1800–4000 m.

3 R **REDHEADED BULLFINCH,** *Pyrrhula erythrocephala* 248
Sparrow +. Rump and vent white. Wings purplish black with large ashy patch and wing-bar. 1000–3900 m.

4 R **GOLDHEADED BLACK FINCH,** *Pyrrhoplectes epauletta* 247
Sparrow. ♀ Yellowish ashy grey and chestnut-brown. A white stripe along wing near back (both sexes). 1400–3900 m.

5 R **SCARLET FINCH,** *Haematospiza sipahi* 246
Bulbul ±. ♀ Brownish olive-yellow. Rump bright yellow. Duars to 2400 m.

6 R **WHITEBROWED ROSEFINCH,** *Carpodacus thura* 244
Sparrow +. Forehead pinkish white. Supercilium pink and white. ♀ Brown, streaked darker. Broad whitish supercilium; rump yellow. 2000–4200 m.

7 R **LARGE ROSEFINCH,** *Carpodacus edwardsii* 245
Sparrow +. Broad pink supercilium. Underparts with dark shaft-streaks. ♀ Underparts buff, finely streaked. 2000–4000 m.

8 R **NEPAL DARK ROSEFINCH,** *Carpodacus nipalensis* 243
Sparrow. Wide maroon breast-band. No red on rump. ♀ Underparts unstreaked olive-brown. 1800–4200 m.

9 R **REDBROWED FINCH,** *Callacanthis burtoni* 241
Sparrow +. White-spotted black wings. ♀ Forehead, supercilium and round eye ochre-yellow. Recorded only at *c.* 3000 m (Sikkim, winter).

10 M **TIBETAN SISKIN,** *Carduelis thibetana* 240
Sparrow –. 1000–3000 m.

11 R **PLAINCOLOURED MOUNTAIN FINCH,** *Leucosticte nemoricola* 241
Sparrow. Rump grey. Tail-coverts broadly white-tipped. Tail blackish, markedly forked. 2000–5300 m.

Below, chin and throat scarlet; sides of throat, and breast, jet black; belly and vent white. **Female**. *Above*, grey-brown; short whitish supercilium and eye-ring. Tail blackish brown with white tip. *Below*, no scarlet throat; breast smoky, fading to whitish on belly (subspecies *confusus* without white malar streak, and *tschebaiewi* with white malar streak).

Status, Habitat, etc. Summer visitor, locally common. Breeds between 3900 and 4800 m; winters in the foothills and duars: rhododendron and juniper scrub, and bare screes in the alpine zone (summer); wet scrub and grass jungle (winter). Habits and behaviour typically robin-like (see 397). Food: insects. Call: alarm, *it . . . it . . .* Song, loud, shrill, whistling, like an accentor's or bush chat's, of 3 or 4 seconds' duration.

399 BLUE CHAT *Erithacus brunneus* (Hodgson)   p. 161
Size Sparrow ± ; length 15 cm (6 in.).

Field Characters Male. *Above*, slaty blue. Lores and cheeks black; a conspicuous white supercilium. *Below*, bright chestnut; centre of belly and vent white. **Female**. *Above*, olive-brown. *Below*, whitish, washed with ochraceous on breast and flanks.

Status, Habitat, etc. Summer visitor. Breeds between 1600 and 3300 m; winters mostly in S. India and Sri Lanka. Dense undergrowth of rhododendron, *Berberis*, etc. in oak or conifer forest (summer); dank ravines and wet evergreen undergrowth, coffee plantations, etc. (winter). Terrestrial, forest-haunting, skulking. Forages singly on ground or low bushes. Behaviour and actions typically robin-like. Food: insects. Call: alarm, a guttural *tuck-tuck* and a high-pitched *tsee*. Song, a jumble of rapid trilling notes *jerri-jerri-jerri-quick-quick-quick-quick*, starting softly and growing louder.

400 REDFLANKED BUSH ROBIN *Erithacus cyanurus* (Pallas) p. 161
Size Sparrow ± ; length 15 cm (6 in.).

Field Characters Male. *Above*, dark blue; forehead, supercilium, shoulder-patch, and rump sky blue. *Below*, centre of throat, breast and belly sullied white; sides of throat blue-black; flanks ochraceous. **Female**. *Above*, olive-brown; rump and tail bluish. A pale eye-ring. *Below*, like male but sides of throat grey-brown; flanks orange (subspecies *rufilatus*).

Status, Habitat, etc. Common altitudinal migrant. Breeds between 3000 and 4400 m; winters from 2500 m down to 1500 in the foothills: undergrowth in conifer forest and dwarf rhododendron bushes, etc. Usually single; shy and secretive. Forages on ground as well as in low bushes, hopping about like robin or making short flycatcher-like aerial sorties after prey. Food: insects. Call: alarm, a throaty *tok* and a mournful *pheeou.* Song, short, of 3 or 4 notes, the middle ones in a lower key.

401 GOLDEN BUSH ROBIN *Erithacus chrysaeus* (Hodgson)   p. 64

Size Sparrow − ; length 15 cm (6 in.).

Field Characters Male. *Above*, crown, back and wings olive-brown; sides of back and rump orange. A yellow supercilium; a black band through eye. Tail orange-and-black with a black terminal band. *Below*, orange, usually with dusky crescent-shaped markings. Female. *Above*, olive; a faint yellowish olive supercilium and buff eye-ring. *Below*, ochre yellow (subspecies *chrysaeus*).

Status, Habitat, etc. Altitudinal migrant. Breeds between 3000 and 4600 m; winters down to the foothills, mostly between 2000 and 1400 m: rhododendron and open conifer forest with stony hillslopes and scrub (summer); dense undergrowth in evergreen forest (winter). A typical robin in habits and behaviour. Food: insects. Call: alarm, a soft distinctive croak *trrr*. Said also to have a beautiful song.

402 WHITEBROWED BUSH ROBIN *Erithacus indicus* (Vieillot)
p. 64

Size Sparrow; length 15 cm (6 in.).

Field Characters Male. *Above*, slaty blue; a very prominent white super-cilium. *Below*, orange-ochraceous. Female. *Above*, olive-brown; a partially concealed white supercilium and pale eye-ring. *Below*, dull rufous-ochre, paler on belly (subspecies *indicus*).

Status, Habitat, etc. Resident, subject to vertical movements; locally common. Breeds between 3000 and 4200 m; winters from 3700 m down to between 3000 and 2000 m. Mixed subalpine forest of birch, fir, rhododendron, etc. (summer); undergrowth of seedlings and bracken in heavy damp forest (winter). Chiefly terrestrial and typically robin-like. Food: insects. Call: a sweet *tuit-tuit*; a croaking *trrr*. Song, a rapid repetition of *tuit-tuit* on an up-and-down scale.

403 RUFOUSBELLIED BUSH ROBIN *Erithacus hyperythrus* (Blyth)
p. 161

Size Sparrow − ; length 12 cm (5 in.).

Field Characters Male. *Above*, deep purplish blue; forehead, supercilium, shoulders and rump bright blue. *Below*, bright chestnut; vent white. Female. *Above*, dark olive-brown; rump slaty blue. Tail bluish black. *Below*, chestnut, browner on breast; centre of belly, and vent, white.

Status, Habitat, etc. Altitudinal migrant, scarce. Breeds between 3400 and 3800 m; winters from about 3500 m down through the foothills. In summer inhabits the dwarf rhododendron zone; in winter forest glades and edges, especially along streams. Typical robin, but habits little known specifically. Food: insects. Call: unrecorded.

404 MAGPIE-ROBIN *Copsychus saularis* (Linnaeus)   p. 53

Size Bulbul; length 20 cm (8 in.).

Field Characters A longish-tailed black-and-white robin. Male. *Above*, glossy blue-black. Wings with a prominent white patch. Tail graduated, black with the outer feathers white. *Below*, throat and breast blue-black; rest white. Female, like male but slaty above, and grey on throat and breast (subspecies *saularis* and *erimelas*).

Status, Habitat, etc. Resident, common. Low country, duars, foothills and up to 1900 m in dry deciduous forest and secondary jungle, often in the neighbourhood of habitations—gardens, orchards, village groves, etc. Singly or pairs. Feeds largely on ground, hopping about with an upright carriage, tail cocked, expanded and jerked upward from time to time, wings partly drooped at sides. In the undulating flight tail constantly flicked open and shut, flashing the white rectrices. Food: mainly insects. Call: a plaintive *swee-ee*; alarm, a harsh *chr-r*. Song, a spirited, clear, thin whistling with short phrases repeated again and again in various combinations.

405 SHAMA *Copsychus malabaricus* (Scopoli)   p. 161

Size Bulbul, with a long tail; length 25 cm (10 in.).

Field Characters Forest counterpart of Magpie-Robin. Male. *Above*, head and back glossy black; rump white. Tail strongly graduated; outer rectrices white, very conspicuous in flight; central rectrices black, elongated. *Below*, throat and breast glossy black; belly and vent rufous. Female similar but black replaced by grey; underparts paler rufous; tail shorter (subspecies *indicus*). In ill-lit forest interior at dusk, the flashing white rump and tail-feathers provide a good diagnostic clue.

Status, Habitat, etc. Resident, rather local—low country, duars and foothills up to 500 m: undergrowth in mixed bamboo forest, and secondary jungle. Distinctly crepuscular and more retiring, but otherwise habits and behaviour similar to Magpie-Robin's. Food: mainly insects. Call: alarm, a harsh scolding; song, spasmodic runs of loud fluty melodious phrases of magpie-robin pattern but richer and of nearer thrush quality.

406 BLUEHEADED REDSTART

Phoenicurus caeruleocephalus (Vigors)   p. 170

Size Sparrow ± ; length 15 cm (6 in.).

Field Characters Male. *Above*, crown and nape bluish ashy; rest including tail and wings black, the latter with a large white patch. *Below*, throat and breast black, belly white. In winter the bluish and black parts get masked by pale brown fringes. Female. *Above* brown; a pale eye-ring. Wing-coverts and tertials with broad whitish edges and tips. Upper tail-coverts and rump rufous. Tail brown, edged with rufous. *Below*, pale fulvous brown fading to white on belly and vent.

Status, Habitat, etc. Altitudinal migrant, uncommon. Breeds between 2400 and 3900 m; winters from 3500 m down to the foothills (1200 m): juniper and open pine forest (summer); opener nullahs and hillsides (winter). Habits as of true redstarts. Keeps singly, descends to ground from bushtop to pick up a titbit, and quivers tail. Food: insects. Call: alarm, a continual plaintive piping *tit, tit* ... Song, reminiscent of Rock Bunting's (533), q.v.

407 **BLACK REDSTART** *Phoenicurus ochruros* (S. G. Gmelin)   p. 161
Size Sparrow ± ; length 15 cm (6 in.).

Field Characters Male. *Above*, crown and back black; wings brown; rump and tail rufous. *Below*, throat and breast black; rest rufous. In winter black parts fringed with grey. Female. *Above*, pale brown with paler eye-ring. Tail rufous with dark brown central feathers. *Below*, pale fulvous brown (subspecies *rufiventris*).

Status, Habitat, etc. Summer visitor. Breeds between 3300 and 5200 m; winters in the foothills, duars and adjacent plains. Boulder-strewn Tibetan steppe facies (summer); open sparsely scrubbed country and about cultivation (winter). Singly or pairs seasonally. Perches on some low eminence, shivers tail constantly and bobs up and down; darts momentarily to ground to pick up a titbit and back to the perch. Food: insects. Call: a mousy *whit* ... *whit* ... *whit* ... like the turning of an unoiled bicycle wheel. Song, a loud pleasing trill of 6 or 7 notes followed by a wheezy jingle.

408 **HODGSON'S REDSTART** *Phoenicurus hodgsoni* (Moore)   p. 161
Size Sparrow ± ; length 15 cm (6 in.).

Field Characters Male. *Above*, forehead to ear-coverts black; crown, nape and mantle ashy grey, whitish on forecrown and supercilium. Wings brown with a prominent white patch; rump and tail rufous, central rectrices dark brown. *Below*, throat and breast black; rest rufous. Female, greyish brown; no wing-patch; rump and tail as in male. *Below*, throat and breast pale grey-brown fading to whitish on belly.

Status, Habitat, etc. Winter visitor, locally common; from 2800 m down through the foothills and duars. May breed in northernmost Arunachal.

Fond of partly dry river-beds and open scrub jungle. Singly or separated pairs in winter; habits and behaviour typical of redstarts. Food: insects, berries. Call: a single clicking *prit* like Bluefronted (409). Alarm, a rattling *trrr*, *tschrrr*. Song, short and tinny.

409 BLUEFRONTED REDSTART *Phoenicurus frontalis* (Vigors)

p. 161

Size Sparrow ± ; length 15 cm (6 in.).

Field Characters Male. *Above*, forehead and supercilium bright blue becoming darker blue on crown and back; rump rufous. Wings brown. Tail rufous with the central rectrices and a broad terminal band blackish. *Below*, throat dark blue continuing as a bib on breast; rest orange-chestnut. In winter crown and back fringed with rusty brown, and tertials edged with buff. Female. *Above*, dark olive-brown; a pale buffish eye-ring; tail pattern as in male diagnostic. *Below*, throat and breast olive-brown; rest orange-brown.

Status, Habitat, etc. Common altitudinal migrant. Breeds between 3000 and 4500 m; winters from 2700 m down to 1000 m. Dwarf juniper and birch scrub above timber-line (summer); open forest, cultivation clearings, pastures, etc. (winter). Typical redstart. Occasionally hawks insects in flycatcher style. Tail wagged up and down more like Whitecapped Redstart (427) than shivered like Black (407). Food: insects, berries, seeds. Call: alarm, a low oft-repeated *ee-tit-tit*. Song, a series of sweet warbling and harsh grating notes—similar to Black Redstart's but less wheezy.

410 WHITETHROATED REDSTART *Phoenicurus schisticeps* (Gray)

p. 129

Size Sparrow ± ; length 15 cm (6 in.).

Field Characters Male. *Above*, crown dull indigo-blue, brighter on forehead; back black with some chestnut on scapulars. Wings blackish with a white patch particularly conspicuous in flight. Rump and upper tail-coverts chestnut. Tail black, base of outer rectrices chestnut. *Below*, chin and throat black with a distinctive large white patch in centre; rest chestnut, whitish on belly. Female. *Above*, head and back brown; rump chestnut. Tail as in male but chestnut replaced by rufous. Wings brown with a small ashy patch. *Below*, a large white throat-patch; rest brown, the feathers fringed with ashy; belly whitish (subspecies *schisticeps*).

Status, Habitat, etc. A high-elevation redstart. Resident, subject to slight vertical movements. Breeds between 2700 and 4500 m; recorded in winter from 4200 down to 1400 m: open park-like forest with scrub oak, dwarf juniper, etc. in summer; open meadows, scrub-covered stony hillsides and dry watercourses in winter. Singles or separated pairs. Typical redstart in habits and behaviour. Food: insects, berries, seeds. Call: alarm, a long-drawn *zieh* followed by a rattling note. Song unrecorded.

411 **DAURIAN REDSTART** *Phoenicurus auroreus* (Pallas) p. 161
Size Sparrow ± ; length 15 cm (6 in.).
Field Characters Male. *Above*, crown, nape and upper back slaty grey; centre of back black; rump and tail rufous, central rectrices blackish. Wings black, with a prominent white patch. *Below*, sides of neck, and throat, black; rest rufous. In winter grey and black parts fringed with brown. Female. *Above*, olive-brown, rump and tail rufous; a buff eye-ring and small whitish wing-patch. *Below*, fulvous brown, throat paler; centre of belly whitish; flanks, vent and under tail-coverts ochraceous (subspecies *leucopterus*).
Status, Habitat, etc. Summer visitor to Arunachal between 2800 and 3700 m. Winters in the foothills west to Sikkim: open country near habitation, secondary jungle, tea gardens, etc. Solitary or pairs. A typical redstart. Food: insects, berries. Call: alarm, *teck, teck*. Song undescribed. Very silent in winter.

412 **GÜLDENSTÄDT'S REDSTART**
Phoenicurus erythrogaster (Güldenstädt) p. 129
Size Sparrow + ; length 16 cm (6 in.).
Field Characters A hardy high-elevation redstart. Male. *Above*, forehead and round eye black; crown and nape white; back and wings black; a large white wing-patch. Rump and tail chestnut. *Below*, throat and breast black; rest chestnut. In winter, feathers of white and black portions fringed with ashy. Female. *Above*, pale brown; a whitish eye-ring. Rump rufous; tail brownish rufous. No wing-patch. *Below*, pale fulvous brown; lower breast and flanks ochraceous buff; centre of belly whitish (subspecies *grandis*).
Status, Habitat, etc. Resident, subject to vertical movements. Breeds between 3600 and 5200 m; winters from 4800 down to 1500 m. Dry stream-beds and boulder-strewn alpine meadows (summer); rocky sparsely scrubbed hillsides and thickets in valley bottoms (winter). Habits and behaviour typical of the genus. Food: insects, berries. Call: not recorded. Song, short, clear, given from perch or in display flight.

413 **PLUMBEOUS REDSTART** *Rhyacornis fuliginosus* (Vigors)

p. 161

Size Sparrow − ; length 12 cm (5 in.).

Field Characters Male. Bluish slaty with chestnut tail and rufous vent. Female. *Above*, dark grey-brown with 2 white wing-bars (of spots) and a pale eye-ring. Tail largely white, with a broad brown terminal band. *Below*, mottled slaty and white (subspecies *fuliginosus*).

Status, Habitat, etc. Common resident, subject to altitudinal movements. Breeds from 4000 m down to 1000; winters through the duars and adjacent low country. Affects clear rushing torrents and rocky streams. Solitary or pairs. Makes short fly-catching sorties from rocks amidstream, or snatches aquatic prey as it drifts past. Continuously opens and shuts tail scissors-wise, and wags it up and down. Food: insects, occasionally berries. Call: a sharp *ziet, ziet*; song, a shrill creaky metallic jingle of 4 or 5 seconds duration, rising in pitch, very reminiscent of a cricket's chirping.

414 **WHITEBELLIED REDSTART** *Hodgsonius phoenicuroides* (Gray)

p. 161

Size Bulbul; length 19 cm (7½ in.).

Field Characters Male. *Above*, slate-blue with two white spots on wing conspicuous in flight. Tail long, graduated, black with a distinctive rufous patch on each side at base of outer rectrices. *Below*, throat and breast slate-blue; belly white. Female. *Above*, brown. Tail as in male, also with rufous patches. *Below*, fulvous; centre of belly whitish (subspecies *phoenicuroides*).

Status, Habitat, etc. Altitudinal migrant. Breeds between 2400 and 3900 m; winters from 1500 m down through the foothills. Thickets of birch, juniper, etc. near timber-line (summer); bush jungle and undergrowth at forest edge (winter). Habits robin-like. Often expands and erects tail vertically, displaying the rufous base. Feeds on ground; makes short flights from bush to bush. Food: insects, berries. Call: alarm, a robin-like *tsiep tsiep tek tek*; song, a loud, rather melancholy whistle of 3 notes *pe-pee-pit*, the middle one longest and highest in pitch.

415 **WHITETAILED BLUE ROBIN** *Cinclidium leucurum* (Hodgson)

p. 161

Size Bulbul − ; length 17 cm (6½ in.).

Field Characters Male. *Above*, indigo blue. Sides of head black; forehead, short supercilium and shoulders brighter blue. Tail blackish with 2 white patches at base, particularly conspicuous in flight. *Below*, throat and breast black; a concealed white patch on each side of neck. Belly dark indigo blue. Female. *Above*, rufous olive-brown. Tail dark brown with white patches as in male. *Below*, throat ochraceous; rest rufous brown, more greyish on belly.

Status, Habitat, etc. Resident, locally common, subject to vertical movements. Breeds between 1200 and 2700 m; winters from 1500 m down through the foothills and duars: shady undergrowth in evergreen forest. Usually solitary and very secretive. Feeds in bushes and on ground. Constantly opens and shuts tail. Habits and behaviour of typical robin. Food: insects, berries. Call: song, of several clear, liquid phrases of robin quality, each given separately with pauses between.

416 BLUEFRONTED ROBIN *Cinclidium frontale* Blyth   p. 176
Size Bulbul ± ; length 19 cm (7½ in.).
Field Characters Very similar sex for sex, and easily mistakable for White-tailed Blue Robin (415); but graduated tail without the white basal patches diagnostic (subspecies *frontale*).
Status, Habitat, etc. Rare and little known. Has been taken at *c*. 2250 m in Sikkim and Darjeeling district, in subtropical wet forest. Habits robin-like, largely ground-feeding. Food and Call: unrecorded.

417 GRANDALA *Grandala coelicolor* Hodgson   p. 176
Size Bulbul + ; length 23 cm (9 in.).
Field Characters Male. Entirely bright purple-blue with a silky sheen, except for black wings and tail. Female. *Above*, brown; head and neck streaked with whitish. Rump tinged with blue. A white wing-patch conspicuous in flight. *Below*, brown, streaked with whitish, more heavily on throat and breast.
Status, Habitat, etc. Resident at high elevations, subject to vertical movements. Locally abundant. In summer between 3900 and 5400 m; in winter between 4300 and 3000 m, and down to 2000 m in very cold weather. Boulder-strewn alpine meadows and rocky slopes above dwarf scrub zone (summer); rocky mountainsides and ridges (winter). Loose flocks of 5 to 50; in winter large restless swarms of several hundred. Flight reminiscent of starlings; other habits of rock thrushes. Feeds mostly on ground hopping about, frequently flicking open wings and tail. Food: insects, berries. Call: *tji-u* or *tju-ti*. Song, merely a quickly repeated variant of the call-note, soft and clear but audible only at close range. Bird very silent on the whole.

418 LITTLE FORKTAIL *Enicurus scouleri* Vigors   p. 177
Size Sparrow − ; length 12 cm (5 in.).

Field Characters *Above*, forehead white; rest of head, neck and upper back black. A conspicuous white triangular bar across wing. Lower back white, interrupted on rump by a black band. Tail short, blackish, slightly forked, outer rectrices white. *Below*, throat black; rest white, smeared on breast and flanks with black. Legs fleshy white. Sexes alike.

Status, Habitat, etc. Resident, locally common, subject to vertical movements. Breeds chiefly between 1800 and 3300 m; winters between 2000 and 1000 m, locally down to 300 m: torrential rocky streams and waterfalls. Ecology similar to Plumbeous Redstart's (413), q.v. Incessantly wags stumpy tail and rapidly opens and shuts it in a rhythmic scissors-like motion. Picks its food off wet rocks; occasionally making short fly-catching sorties after midges etc. Food: chiefly aquatic insects. Call: unrecorded.

419 BLACKBACKED FORKTAIL *Enicurus immaculatus* (Hodgson)
p. 177

Size Bulbul, with long tail; length 25 cm (10 in.).

Field Characters General effect of a pied wagtail. *Above*, forehead and supercilium white; rest of head and back black. Rump white continued as a white band across black wings. Tips of secondaries white. Tail long, very deeply forked and graduated; outer rectrices white, inner black with white tips. *Below*, throat black, rest white. Sexes alike.

Status, Habitat, etc. Resident: foothills, normally below 1450 m down through the duars—on rocky hill streams and fast-flowing rivers in dense damp forest. Usually solitary, tripping lightly over stones in water or on bank, incessantly wagging tail. Flight and behaviour very reminiscent of wagtail. Food: insects. Call: a sharp *curt-see*, the second syllable prolonged and much higher.

420 SLATYBACKED FORKTAIL *Enicurus schistaceus* (Hodgson)
p. 177

Size Bulbul, with long tail; length 25 cm (10 in.).

Field Characters *Above*, narrow forehead and short supercilium white; crown and back slaty; rump white. Wings black with a broad white band across, a small white patch on primaries, and white tips to tertials. Tail as in Blackbacked (419). *Below*, chin and sides of throat black; rest white, slaty on sides of breast. Sexes alike.

Status, Habitat, etc. Resident, subject to vertical movements. Breeds between 300 and 1600 m; in winter extends down through the duars into the adjacent plains. Habitat, ecology and food as of 419. Call: a high-pitched screechy single note.

421 LESCHENAULT'S FORKTAIL *Enicurus leschenaulti* (Vieillot)
Size Myna, with long tail; length 30 cm (12 in.).
Field Characters The largest and blackest of the forktails. Very similar to Blackbacked (419) but larger. White of forehead extends to crown (not only as a supercilium); black of underparts extends to breast and flanks (subspecies *indicus*). Sexes alike.
Status, Habitat, etc. Resident, subject to seasonal movements. A low-elevation forktail found in the duars and some distance out in the adjacent plains, at fast-flowing streams and rivulets in dense evergreen forest. Habits, ecology and behaviour as of the genus. Call: a very shrill single note well compared to the screech of the stopper sharply turned in the neck of a glass bottle.

422 SPOTTED FORKTAIL *Enicurus maculatus* Vigors   p. 177
Size Bulbul, with long tail; length 25 cm (10 in.).
Field Characters *Above*, forehead and forecrown white; sides of head and nape black. Back black, spotted with white; rump white. A broad white wing-bar; tertials white-tipped. Tail deeply forked and graduated; outer rectrices white, the others black tipped with white. *Below*, throat and breast black; rest white (subspecies *guttatus*). Sexes alike.
Status, Habitat, etc. Resident, common. Breeds between 600 and 3000 m. Affects small streams in dense forest; in winter also river-beds in duars and at base of the hills. Ecology, habits and behaviour as of the genus. Food: aquatic insects and tiny molluscs. Call: a shrill rasping *kreee* or *tseeek* very like a Whistling Thrush's (431), uttered in flight; also a sharp creaky *cheek-chik-chik-chik-chik*, at rest and on the wing.

423 PURPLE COCHOA *Cochoa purpurea* Hodgson   p. 176
Size Myna +; length 30 cm (12 in.).
Field Characters Male. *Above*, crown lavender-blue; supercilium, ear-coverts and nape black. Back brownish purple. Wings purple-lavender with black tip and edge; a pale lavender patch on shoulder and another on middle of wing. Tail purple-lavender with black terminal band. *Below*, uniform brownish purple. Female similar but brownish purple parts replaced by reddish brown above, brownish clay-colour below.
Status, Habitat, etc. Resident, rare. Recorded between 1000 and 3000 m: dense humid evergreen forest, undergrowth in ravines, etc. Singles or pairs. Very quiet and secretive. Feeds on ground and in trees. Food: berries, insects, molluscs. Call: only a low chuckle recorded.

424 GREEN COCHOA *Cochoa viridis* Hodgson   p. 176
Size Myna +; length 30 cm (12 in.).

Field Characters Male. *Above*, crown to nape sky-blue; lores and super-cilium black; ear-coverts dark blue. Back and wing-coverts deep green with obsolete black bars; rump deep green. Wings black with a broad pale blue band, a narrow black line across it, and a small black patch. Tail blue, outer rectrices and terminal band black. *Below*, deep green, washed with blue on throat and belly. **Female** similar but greater wing-coverts and secondaries marked with yellowish brown instead of blue.

Status, Habitat, etc. Resident, rare. Breeds between 700 and 1500 m. Affects undergrowth in dense evergreen forest. Habits and behaviour like 423, but little known. **Food:** berries, insects, molluscs. **Call:** 'a harsh note' only recorded.

425 COLLARED BUSH CHAT *Saxicola torquata* (Linnaeus)   p. 129
Size Sparrow − ; length 13 cm (5 in.).

Field Characters Male. *Above*, black except for white rump and white wing-patch. *Below*, throat black; a large white patch on each side of neck ('the collar'). Breast rufous-chestnut paling to buff on belly. In winter, feathers of upperparts broadly fringed with brown. **Female.** *Above*, rufous-brown streaked darker; rump pale rufous. A white wing-patch. Tail blackish brown. *Below*, pale fulvous, more rufous on breast (subspecies *przevalskii* and *stejnegeri*).

Status, Habitat, etc. Winter visitor: duars and foothills up to 1300 m, exceptionally to 2400 m (Thimphu): high grass, terraced cultivation, pastureland and sparsely scrubbed hillsides. Keeps in pairs. Perches on low bushes, constantly flicking open tail and jerking it up and down. Descends to ground every now and again to pick up some titbit, returning to the same bush or flying on to another. Also 'fly-catches' in air. **Food:** insects. **Call:** alarm, *check-check* or *pee-tack*. Song, a short, lively warble mixed with mimicked calls of other birds.

426 DARK-GREY BUSH CHAT *Saxicola ferrea* Gray   p. 177
Size Sparrow; length 15 cm (6 in.).

Field Characters Male. *Above*, dark ashy grey with black streaks; a white supercilium; sides of head black. Tail black with whitish edges. A concealed white shoulder-patch visible in flight. *Below*, throat white; breast and flanks grey; belly whitish. In winter, feathers of upperparts and breast tipped with rufous-brown. **Female.** *Above*, rufous-brown faintly streaked darker; a pale supercilium; sides of head dark brown; rump rusty; tail blackish edged rufous. *Below*, chin and throat white; rest fulvous, darkest on breast.

Status, Habitat, etc. Resident, subject to short altitudinal movements.

Breeds between 1500 and 3300 m; winters from 2400 down through the foothills, duars and adjacent plains. Open scrub-covered hillsides, edges of forest glades, terraced cultivation, etc. Habits and behaviour similar to Collared (425), typical of bush chats. Food: chiefly insects. Call: alarm, *zee-chunk* repeated at short intervals; also a sharp *tak-tak-tak-tak*. Song, a short pleasant rather feeble trill *sisiri-swirrr*, and variations.

427 WHITECAPPED REDSTART
Chaimarrornis leucocephalus (Vigors)   p. 177
Size Bulbul; length 19 cm (7½ in.).

Field Characters *Above*, crown and nape white; forehead, sides of head, back and wings black. Rump and tail chestnut, the latter with black terminal band. *Below*, throat and breast black; rest rich chestnut. Sexes alike.

Status, Habitat, etc. Common altitudinal migrant. Breeds between 1800 and 5100 m; winters down through the foothills and duars and a short distance out in the adjacent plains. Affects the larger rapid mountain streams; in winter, at low elevations, clear shingly rivers and canals etc. Singles or pairs on projecting rocks or banks in streams. Picks up titbits from water as they float past or by erratic zigzag aerial sorties in flycatcher fashion. Bobs or curtsies on alighting, wagging expanded tail jerkily up and down with the wings drooping at sides. Food: insects, molluscs, berries. Call: alarm, a shrill piercing *tseee*. Song, a long, melodious, somewhat melancholy whistling.

428 BLUEHEADED ROCK THRUSH
Monticola cinclorhynchus (Vigors)   p. 177
Size Bulbul; length 17 cm (6½ in.).

Field Characters Male. A small blue, chestnut and black thrush with a white patch on wings, diagnostic at rest and in flight; back black, rump rufous. *Below*, throat blue; rest orange-rufous. In winter black feathers of back edged with fulvous. Female. *Above*, olive-brown. *Below*, squamated whitish and dark brown.

Status, Habitat, etc. Summer visitor, common. Breeds between 1000 and 2200 m; winters down through the duars, and chiefly in the Indian plains. Open rocky grass-covered hillslopes (summer); secondary jungle and well-wooded country in evergreen or moist-deciduous biotope (winter). Singles or pairs. Largely arboreal, but frequently descends to ground to rummage among mulch for food. Perches bolt upright and slowly wags tail up and down. Food: insects, small lizards, etc., berries, seeds. Call: alarm, a grating *goink-goink*. Song, reminiscent of Collared Bush Chat's but richer and of distinct thrush-quality.

429 CHESTNUTBELLIED ROCK THRUSH *Monticola rufiventris*
(Jardine & Selby) p. 176

Size Myna ± ; length 24 cm (9½ in.).

Field Characters M a l e. *Above*, brilliant blue with some blackish on mantle. Lores, ear-coverts and sides of neck black. *Below*, throat blackish blue; rest chestnut. F e m a l e. *Above*, olive-brown with dark crescent-shaped bars. *Below*, squamated dark brown and buff. Distinguished from similar female Blue Rock Thrush (430) by whitish throat-patch.

Status, Habitat, etc. Resident, fairly common, subject to vertical movements. Breeds between 1200 and 3300 m; winters down through the foothills and duars. Habitat, behaviour and food as in Blueheaded (428). C a l l: alarm, a harsh, rasping rattle *chhrrr*. Song, a pleasant warbling resembling song of Blueheaded, *teetatewleedee-tweet tew*, repeated several times with variations.

430 BLUE ROCK THRUSH *Monticola solitarius* (Linnaeus) p. 177

Size Bulbul + , length 23 cm (9 in.).

Field Characters M a l e. Overall bright blue. Wings and tail brown. In winter, feathers of upperparts fringed with fulvous, breast with brown, belly with white. F e m a l e. *Above*, grey-brown, barred with blackish on rump. A pale wing-bar conspicuous in flight. *Below*, whitish, cross-barred (squamated) with brown. Distinguished from rather similar female Chestnutbellied (429) by less boldly barred underparts and lack of whitish throat-patch (subspecies *pandoo*).

Status, Habitat, etc. Summer visitor. Breeds (presumably) between 1200 and 3000 m; winters down through the foothills, duars and Indian plains. Barren rocky hillsides (summer); broken rocky country, stone quarries, ancient forts and buildings, etc. (winter). Pairs in summer, solitary in winter. Sits bolt upright, at times bowing, curtsying and flirting tail vehemently. Descends to ground to pick up a morsel and flies back to the same or a different perch. Sometimes 'fly-catches' winged insects. F o o d: insects, small lizards, etc., berries. C a l l: song, a short, typically thrush-like melodious warble given from a perch or on the wing.

431 WHISTLING THRUSH *Myiophonus caeruleus* (Scopoli) p. 177

Size Pigeon; length 35 cm (14 in.).

Field Characters Dark purple-blue spotted with glistening blue. Forehead, shoulder and edge of wing, and tail, brighter blue. A half-dozen white spots on median wing-coverts. (Birds from eastern Arunachal lack these.) Bill yellow (subspecies *temminckii* and *eugenei*). Sexes alike.

Status, Habitat, etc. Resident, common, subject to vertical movements. Breeds normally between 1500 and 4000 m; winters from 2400 down through the foothills and duars to some distance out in the adjacent plains. Affects rocky streams and torrents in wooded ravines and gorges. Singly or pairs. Feeds on ground moving in long hops and turning over dry leaves etc. Behaviour and actions typically blackbird-like, but is more crepuscular. Food: aquatic insects, snails, crabs, etc.; also berries. Call: a strident far-carrying *tzeet tze-tze-tzeet* and a loud, shrill *kreee*. Song, a loud, clear and resonant whistling of remarkably human quality.

432 ORANGEHEADED GROUND THRUSH

Zoothera citrina (Latham) p. 177

Size Myna − ; length 21 cm (8½ in.).

Field Characters Male. Entire head, nape and underparts orange-chestnut. Vent white. Rest of upperparts bluish grey; a white wing-bar; outer tail-feathers lightly tipped white. Female like male but with mantle strongly tinged olive-brown (subspecies *citrina*).

Status, Habitat, etc. Resident and partially migratory. Breeds from 1600 m down to the foothills (and duars?); winters irregularly on the eastern side of the Peninsula and in Sri Lanka. Affects undergrowth in moist-deciduous forest, mixed secondary and bamboo jungle, village groves, etc. Singly or pairs. Rummages quietly for food among leafy débris on the ground under shrubbery. Food: insects, earthworms, berries. Call: alarm, a high-pitched *kreeee*. Song, loud, rich and melodious, of thrush quality, with calls of other birds interlarded.

433 PLAINBACKED MOUNTAIN THRUSH

Zoothera mollissima (Blyth)

Size Myna + ; length 25 cm (10 in.).

Field Characters *Above*, rufescent olive-brown; a conspicuous pale eye-ring. A white wing-patch concealed at rest, prominent in flight. Tips of outer tail-feathers white. *Below*, buff changing to white on belly, boldly marked with blackish crescentic spots (subspecies *mollissima*). Sexes alike. Indistinguishable with certainty from *Z. dixoni* (434) often found side by side.

Status, Habitat, etc. Altitudinal migrant. Breeds between 3000 and 4300 m; winters from 3600 m down to 1300: grassy boulder-strewn hill-slopes with dense bushes and oak and rhododendron forest (summer); open bush country around hill cultivation and in stream valleys, etc.

Plate 34, artist D. M. Henry

PARROTBILLS, TITS

1 R BLACKFRONTED PARROTBILL, *Paradoxornis nipalensis* page 136
Sparrow — . Throat black. 1200–3300 m.

2 R FULVOUSFRONTED PARROTBILL, *Paradoxornis fulvifrons* 136
Sparrow — . Throat fulvous. 2700–3400 m.

3 R SULTAN TIT, *Melanochlora sultanea* 218
Bulbul. 300–1800 m.

4 R LESSER REDHEADED PARROTBILL, *Paradoxornis atrosuperciliaris* 136
Sparrow. Crown rufous; ear-coverts not black. Up to 1500 m.

5 R YELLOWBROWED TIT, *Sylviparus modestus* 220
Sparrow — . 1500–4000 m.

6 R COAL TIT, *Parus ater* 219
Sparrow — . Cheeks and nuchal patch white. Two white wing-bars (rows of spots). 1800–3600 m.

7 R BROWN CRESTED TIT, *Parus dichrous* 219
Sparrow — . From 2200 m up to timber-line.

8 R RUFOUSFRONTED TIT, *Aegithalos iouschistos* 221
Sparrow — . Chin blackish; centre of throat silvery white. 2400–3600 m.

0   3   6   9 cm

Winton Cneado

Plate 35, artist Winston Creado

NUTHATCHES, FLOWERPECKERS, RUBYCHEEK, TREE CREEPER, WAGTAILS, PIPITS

(winter). Usually pairs; very shy. Feeds quietly in shady spots on ground; flies up into overhanging branches when disturbed. Food: insects, berries, snails, seeds. Call: a rattling alarm-note. Song undescribed.

434 LONGTAILED MOUNTAIN THRUSH
Zoothera dixoni (Seebohm)

Size Myna + ; length 25 cm (10 in.).

Field Characters Very similar to *Z. mollissima* (433), usually distinguishable from it by presence of two buff wing-bars. Sexes alike.

Status, Habitat, etc. Altitudinal migrant. Breeds between 2000 and 4000 m; winters between 2700 and 1500 m: dense fir, birch and rhododendron forest near tree-line (summer); same biotope as Plain-backed (433) in winter. Behaviour, food, etc. of the two very similar. Call: comparable description unavailable.

435 SMALLBILLED MOUNTAIN THRUSH
Zoothera dauma (Latham)   p. 176

Size Myna + ; length 25 cm (10 in.).

Field Characters *Above*, olive-brown with buff and black crescent-shaped markings. Closed wings with buff and blackish bars. A large buff underwing-patch prominent in flight. *Below*, throat and belly white; breast and flanks buff with bold blackish crescentic spots (subspecies *dauma*). Sexes alike.

Status, Habitat, etc. Altitudinal migrant. Breeds between 2400 and 3400 m; winters from 1800 m down through the foothills, duars and adjacent plains: heavy oak and silver fir forest (summer); dense forest with grassy clearings, bracken or seedling undergrowth, well-wooded stream banks, bamboo jungle, etc. (winter). Solos or pairs. Feeds on ground, flying silently up into trees when disturbed. Food: insects and berries. Call: song, loud, rich, typically thrush-like, of several disconnected phrases with long pauses between.

436 LARGE BROWN THRUSH *Zoothera monticola* Vigors   p. 53

Size Myna + ; length 30 cm (12 in.).

Field Characters A stout, short-tailed dark thrush with very large curved bill. *Above*, dark slaty brown. *Below*, throat whitish, sparsely spotted with blackish; breast and flanks olive-brown, the breast with blackish and buff spots; belly white, spotted with blackish. A large underwing-patch prominent in flight (subspecies *monticola*). Sexes alike.

Status, Habitat, etc. Resident, scarce, subject to vertical movements. Breeds between 2000 and 3000 m; winters down through the foothills and duars. Affects small mountain streams running through dense damp forest, and matted undergrowth on swampy ground. Solos. Somewhat crepuscular. Feeds on the damp forest floor and in stream-beds, tossing aside fallen leaves and digging into soft earth with the powerful bill. Food: insects, snails, berries. Call: a beautiful mellow whistle (not adequately described).

437 LESSER BROWN THRUSH *Zoothera marginata* Blyth

Size Myna + ; length 25 cm (10 in.).

Field Characters A stout, short-tailed dark thrush, like 436 but smaller, also with large curved bill. *Above*, rufescent olive-brown. *Below*, throat whitish, bounded on each side by a blackish streak; breast and belly buffish white scalloped with olive-brown; flanks olive-brown with whitish streaks. A buff underwing-patch prominent in flight. Sexes alike.

Status, Habitat, etc. Resident, uncommon, subject to vertical movements. Breeds (presumably) around 2000 m; in winter found in the lower foothills and duars. Affects small watercourses in damp forest. Habits and food similar to 436. Call: unrecorded.

438 WHITECOLLARED BLACKBIRD *Turdus albocinctus* Royle

Size Myna + ; length 25 cm (10 in.).

Field Characters M a l e. Entirely black with white throat and a broad white collar round neck extending to upper back. Bill, legs and feet yellow. Female rufous-brown, paler below and scalloped; collar dull ashy instead of white.

Status, Habitat, etc. Resident, subject to vertical movements. Breeds between 2700 and 4000 m; winters from 3000 m down to the duars and base of the hills. Conifer, oak and rhododendron forest in summer. Solos or pairs. Feeds on the ground as well as up in trees. Food: insects, drupes and berries. Call: alarm, the characteristic throaty *tuck-tuck-tuck-tuck* of thrushes. Song, rich, mellow and varied—of several bursts of notes on a descending scale.

439 GREYWINGED BLACKBIRD *Turdus boulboul* (Latham)   p. 192

Size Myna + ; length 30 cm (12 in.).

Field Characters M a l e. Overall black with a large and distinctive pale grey wing-patch. Feathers of belly and vent with whitish fringes. Eye-rim yellow, bill orange. Female brownish ashy tinged with olivaceous; a rufous-brown wing-patch instead of male's grey.

Status, Habitat, etc. Resident, subject to vertical movements. Breeds between 1800 and 3000 m; winters from 1800 m down to the foothills and duars. Humid oak and rhododendron forest (summer); secondary and bush jungle and village precincts (winter). Solo or pairs; small flocks in winter. A typical thrush in habits and behaviour. Food: insects, earthworms, fruits and berries. Call: alarm, the typical blackbird chuckles *chŭk, chŭk, chŭk*. Song, rich, fluty, far-carrying: of remarkable variety and mellowness and perhaps one of our finest bird songs.

440 BLACKBIRD *Turdus merula* Linnaeus

Size Myna + ; length 25 cm (10 in.).

Field Characters Male entirely black with yellow legs and orange bill. Female dark brown with brown legs and bill (subspecies *maximus*).

Status, Habitat, etc. Resident, subject to vertical movements. Breeds between 3600 and 4500 m, descending to 2000 in winter. Dwarf juniper and rhododendron scrub on steep rocky slopes above timber-line, and alpine meadows in summer. Solos or pairs; small flocks in winter. A typical thrush in habits and behaviour. Food: insects, fruits, berries, small lizards, etc. Call: the characteristic throaty alarm-notes *chut-ut-ut*. Song, loud, rich, fluty, ending in creaky chuckling notes.

441 GREYHEADED THRUSH *Turdus rubrocanus* Hodgson

Size Myna + ; length 25 cm (10 in.).

Field Characters Male. *Above*, head and neck creamy grey; rest of upperparts chestnut. Wings and tail blackish. *Below*, throat and upper breast whitish; upper belly and flanks chestnut; centre of belly whitish, sometimes with dusky spots. Female much duller: head darker and brownish; chestnut paler, scalloped on underparts with pale greyish (subspecies *rubrocanus*).

Status, Habitat, etc. Resident between 2000 and 3000 m, subject to vertical movements. Fir and horse-chestnut forests in summer, more open country and orchards, etc. in winter. Solos or pairs; small flocks in winter, often mixed with other thrushes. Quiet and unobtrusive. Food: insects, berries. Call: song, loud, rich, of several phrases, each reiterated two or three times, recalling that of the European Song Thrush, *T. ericetorum*. Also has the characteristic harsh *chuck-chuck-chuck* of blackbirds.

442 KESSLER'S THRUSH *Turdus kessleri* Przevalski

Size Myna + ; length 25 cm (10 in.).

Field Characters Male. *Above*, head, nape and upper back black; scapulars rufous-brown; centre of back whitish buff changing to rufous-brown on rump. Wings and tail black. *Below*, throat and breast black; upper belly whitish buff changing to chestnut on lower parts; under tail-coverts black, margined with chestnut. Female. *Above*, head, nape, upper back, wings and tail blackish; back grey-brown; rump tinged with tawny. *Below*, throat, sides of head and upper breast blackish; rest grey-brown, tinged with tawny on belly.

Status, Habitat, etc. A rare straggler from E. Tibet and W. China in winter. Has been recorded twice: at 2700 and 3700 m near the northern border (Sikkim) in low juniper scrub and cultivation. Keeps in flocks in winter in company with other *Turdus* thrushes. Food: chiefly juniper berries (winter). Call: loud harsh chuckles, and a soft *dug, dug*.

443 DARK THRUSH *Turdus obscurus* Gmelin   p. 192
Size Myna ± ; length 23 cm (9 in.).
Field Characters Male. *Above*, olive-brown, greyer on nape and sides of head. Lores black; a broad white supercilium; a white spot under eye. *Below*, chin white, throat slaty grey; breast and flanks pale fulvous brown; belly white; under tail-coverts white, edged with brown at base. Female similar, but throat white streaked with brown; ear-coverts streaked with white.

Status, Habitat, etc. Rare winter straggler to the foothills (Sikkim and Arunachal). Keeps in flocks with Blackthroated and other thrushes chiefly in forest. Feeds largely on ground. Food; insects, snails, berries. Call: a thin pipit-like *zip-zip* when taking off.

444 BLACKTHROATED THRUSH
        Turdus ruficollis atrogularis Jarocki   p. 192
Size Myna ± ; length 25 cm (10 in.).
Field Characters Male. *Above*, grey-brown slightly spotted with blackish on crown and nape. Lores, sides of neck and supercilium black. *Below*, throat and breast black (with whitish fringes in winter); rest white. Female. *Above*, brown. *Below*, throat streaked dark brown and whitish; a broad blackish breast-band, with whitish fringes to the feathers. Rest white, with sparse brownish streaks on belly and sides.

Status, Habitat, etc. Winter visitor, fairly abundant; from the plains, duars and foothills up to 4000 m: cultivation, sparsely scrubbed hillsides and pastures. Highly gregarious; usually seen in flocks associated with other thrushes. Food: insects, earthworms, snails, berries. Call: a thin *seet*; alarm, a throaty chuckle *which-which-which*.

445 REDTHROATED THRUSH *Turdus ruficollis* Pallas   p. 192
Size Myna ± ; length 25 cm (10 in.).
Field Characters Male. Like 444 but black of throat and supercilium replaced by chestnut (the feathers with whitish fringes in winter). Tail

rufous except central rectrices. Female variable; mostly like male but the chestnut paler, mottled with white and spotted with black. A moustachial line of dark brown spots.

Rufous outer rectrices completely diagnostic of this subspecies.

Status, Habitat, etc.   Winter visitor, fairly abundant. Habitat, behaviour, food, calls, etc. not different from the Blackthroated subspecies (444).

446 DUSKY THRUSH *Turdus naumanni* Temminck

Size Myna ± ; length 23 cm (9 in.).

Field Characters A dark thrush with conspicuous white supercilium, rufous wings, and blackish breast and flanks with broad white scale-like markings; belly white. **Female**. *Above*, grey-brown, more rufous on rump. Wings rufous as in male. *Below*, throat and sides of neck pale cream speckled with dark brown. Breast and flanks whitish with scale-like chestnut markings; belly white (subspecies *eunomus*).

Status, Habitat, etc.   Rare winter straggler. Recorded in the Dafla Hills of Arunachal Pradesh: thinly wooded country, open fields and grasslands. Ecology as of other thrushes. Call: a starling-like *spirr*; alarm, a rapid *kveveg*.

WRENS: Troglodytidae

447 WREN *Troglodytes troglodytes* (Linnaeus)   p. 53

Size Sparrow − ; length 9 cm (3½ in.).

Field Characters A tiny dark brown bird with short erect tail. *Above*, rufous-brown, narrowly barred with blackish on wings, rump and tail. A pale eye-ring. *Below*, paler, more closely barred; somewhat whitish on belly (subspecies *nipalensis*). Sexes alike.

Status, Habitat, etc.   Resident, common, subject to vertical movements. Breeds between 2700 and 4700 m; descends lower in winter, sometimes to 2200 m. Affects rock and boulder country in fir, birch and rhododendron forest near timber-line, and higher in the dwarf juniper zone. Solitary, restless and inquisitive. Creeps mouse-like through brushwood and rock crevices, with stumpy tail cocked and wings partially drooped. Food: insects. Call: alarm, a harsh scolding *ter-tzer-tzrrrr*. Song, a spirited high-pitched jumble of rapid vibrant notes, astonishingly loud for the bird's size.

DIPPERS: Cinclidae

448 WHITEBREASTED DIPPER *Cinclus cinclus* (Linnaeus) p. 192
Size Myna ± ; length 20 cm (8 in.).
Field Characters A dumpy short-tailed thrush-like bird of torrential
streams. *Above*, head and upper back chocolate-brown; rest slaty, with
brown scale-like markings on back and rump. *Below*, throat and breast
white; belly chocolate-brown, the feathers faintly tipped white (subspecies *przewalskii*). Sexes alike.
Status, Habitat, etc. Resident; common. Breeds between 3600 and
4700 m; winters down to 2700. Affects swift-flowing torrents, and
glacial tarns. Solitary or widely separated pairs. Jumps from rock to
rock amid stream, bowing and curtsying with a bend-stretch of legs
and cocking of stub tail. Feeds under water by plunging from low rock.
Flight swift with rapid hovering wing action. Food: aquatic insects.
Call: a shrill *dzit, dzit* in flight. Song, wren-like—lively, high-pitched
and piercing.

449 BROWN DIPPER *Cinclus pallasii* Temminck p. 192
Size Myna ± ; length 20 cm (8 in.).
Field Characters A squat, stub-tailed chocolate-brown thrush-like bird
of clear rocky mountain streams (subspecies *dorjei*). Sexes alike.
Status, Habitat, etc. Common resident, subject to vertical movements.
Breeds between 1000 and 4200 m elevation; reaches the lower foothills
in non-breeding season. On rapid rocky mountain streams. Occupies a
lower altitudinal zone than Whitebreasted; ecology, habits, food and
calls of both very similar.

ACCENTORS or HEDGE-SPARROWS: Prunellidae

450 ALPINE ACCENTOR *Prunella collaris* (Scopoli) p. 192
Size Sparrow; length 17 cm (6½ in.).
Field Characters *Above*, head greyish brown; back streaked with dark
brown; rump rufescent. Wing-coverts blackish, their white tips forming
two wing-bars. Tail-feathers blackish, tipped with white. *Below*,
chin and centre of throat white, finely barred with brown. Sides of
throat, breast and belly grey; flanks rusty, white-tipped posteriorly.
Under tail-coverts blackish edged with white (subspecies *nipalensis*).
Sexes alike.

Status, Habitat, etc. Resident, common, subject to vertical movements. Breeds from 3700 to over 5500 m; winters between 4800 and 2400 m or slightly lower. Affects rock-strewn meadows, cliffs, screes, and open ridges; in winter often around upland villages. Pairs or small flocks. Hops about quietly on the ground or on boulders in search of food. Food: insects, seeds. Call: a rippling *chirriririp* when flushed. Song, a pleasing warble of lark-quality given from a rock or in flight.

451 ALTAI ACCENTOR *Prunella himalayana* (Blyth)   p. 192

Size Sparrow; length 15 cm (6 in.).

Field Characters Distinguished from rather similar Alpine Accentor (450) by white gorget on chin and throat, rufous-and-white striped breast (*v.* grey), and pale supercilium. Sexes alike.

Status, Habitat, etc. Winter visitor between 2500 and 4000 m elevation, keeping to bare rocky hillsides. More gregarious than Alpine Accentor; often in largish close-packed flocks in company with Mountain Finches, feeding quietly on stony ground. Flight swift and finch-like. Food: insects, seeds. Call: a low, silvery finch-like twitter while feeding and also on the wing.

452 ROBIN ACCENTOR

Prunella rubeculoides (Horsfield & Moore)   p. 192

Size Sparrow; length 17 cm (6½ in.).

Field Characters *Above*, head brownish grey; rest pale brown with darker streaks on back; two whitish wing-bars. *Below*, throat brownish grey; breast rufous; belly pale cream; lower flanks streaked with rufous-brown. Sexes alike.

Status, Habitat, etc. Resident, common, subject to vertical movements. Breeds between 3600 and 5300 m; descends in winter to 1200 m. Dwarf willows and furze patches near streams, and low scrub in valley bottoms (summer); stony rocky ground, and often in and around upland villages (winter). In small flocks in winter, hopping quietly and feeding on the ground. Food: insects, seeds. Call: alarm, a bunting-like *zieh-zieh*. A pleasant but feeble (song?) *tililili*.

453 RUFOUSBREASTED ACCENTOR *Prunella strophiata* (Blyth)

p. 193

Size Sparrow; length 15 cm (6 in.).

Field Characters *Above* dark-streaked rufous-brown; a prominent orange-rufous supercilium. *Below*, throat whitish, streaked with black; breast rufous (subspecies *strophiata*). Sexes alike.

Distinguished from Robin Accentor (452) by streaked throat and belly and prominent supercilium; from Altai Accentor (451) by rufous (*v.* pale grey) supercilium, streaked whitish throat, unstreaked rufous breast, and dark streaks on belly.

Status, Habitat, etc.   Resident, common, subject to vertical movements. Breeds between 3400 and 4300 m; winters from 3600 down to 1300 m or so. Rhododendron forest near timber-line, dwarf juniper, etc. above this limit (summer); open scrub and bushes on fallow land often around villages (winter). A great skulker, preferring to thread its way through the rootstocks and tangles of branches rather than fly. Takes hurried hedge-hopping flights when flushed. F o o d : insects, seeds. C a l l : alarm, a high-pitched chattering *tir-r-r*. Song, very wren-like, less loud.

454 **BROWN ACCENTOR** *Prunella fulvescens* (Severtzov)   p. 192
Size Sparrow; length 15 cm (6 in.).
Field Characters *Above*, pale brown, streaked darker on back; a prominent long white supercilium and blackish 'cheeks'. Two faint whitish wing-bars. *Below*, buffish, more ochraceous on breast (subspecies *sushkini*). Sexes alike.
Status, Habitat, etc.   Rare straggler from Tibet. Has occurred in Sikkim, and may be expected in Bhutan and Arunachal. In winter, frequents low scrub on stony hillsides, often around villages. Less secretive than Rufousbreasted (453), commonly feeding in the open and perching on walls and roofs. F o o d : insects, seeds. C a l l : a bunting-like *ziet, ziet*.

455 **MAROONBACKED ACCENTOR** *Prunella immaculata*
     (Hodgson)   p. 192
Size Sparrow; length 15 cm (6 in.).
Field Characters A dark slaty-grey accentor with chestnut wings, rufous vent and conspicuous pale yellow eyes. Back and rump rufescent olive-brown and maroon. Sexes alike.
Status, Habitat, etc.   Scarce resident (?), subject to vertical movements. Breeds (in SE. Tibet) between 2900 and 4200 m; winters between 3700 and 2100 m. Humid, mossy conifer and rhododendron forest (summer); secondary jungle, forest edges along terraced fields, boggy nullahs, etc. (winter). More secretive even than Rufousbreasted. Feeds quietly on the ground under dense brushwood, seldom venturing into the open. Swiftly 'hedge-hops' and disappears when flushed. F o o d : insects, seeds. C a l l : a metallic *zieh-dzit*.

TITMICE: Paridae

456 SULTAN TIT *Melanochlora sultanea* (Hodgson)   p. 208
Size Bulbul; length 20 cm (8 in.).
Field Characters A yellow-crested black-and-yellow arboreal bird,
reminiscent of a bulbul (*Pycnonotus*). **Male.** *Above*, crown and crest
bright yellow; rest metallic black. *Below*, throat black; rest bright
yellow. **Female** similar but back blackish olive, throat yellowish
olive (subspecies *sultanea*).
Status, Habitat, etc. Resident, common, between 300 and 1000 m,
locally higher (up to 1800 m). Open foothills forest, deciduous and
evergreen, with large trees; often near cultivation. Pairs or small parties.
Actions and behaviour essentially tit-like. Hunts for insects among the
foliage, clinging to boughs and sprigs upside down and in all manner of
acrobatic positions. Has a peculiar hovering or volplaning flight from
branch to branch. **Food:** insects, berries, seeds. **Call:** contact, a shrill
chip-tree-trr while foraging in company. Alarm, a harsh rolling *krikrew*.
Song, a musical jingling *tew-r-r* normally repeated three times,
punctuated by harsh *chur-chur* or *chuchuk*.

457 GREY TIT *Parus major* Linnaeus
Size Sparrow ± ; length 13 cm (5 in.).
Field Characters *Above*, crown black; cheeks white; back grey; a whitish
patch on nape. Wings dark brown: tertials broadly edged with pale
ashy; a white wing-bar. Tail blackish with white outer rectrices. *Below*,
throat black, continued as a broad black band down middle of underparts; flanks ashy (subspecies chiefly *nipalensis*). Sexes alike.
Status, Habitat, etc. Fairly common resident in the duars and foothills
up to 900 m; replaced at higher elevations by Greenbacked Tit (458):
light forest, cultivation, gardens, village groves, etc. Pairs or scattered
parties, often associated with other small insectivorous birds. Very
active. Hunts among foliage, clinging to sprigs sideways or upside down,
peering into crannies for caterpillars etc. **Food:** insects, berries,
flower buds, seeds. **Call:** contact, *tsee tsee tsee* while foraging. Song, a
loud, clear, whistled ditty *whee-chichi, whee-chichi, whee-chichi* repeated
several times with a short break before resumption.

458 GREENBACKED TIT *Parus monticolus* Vigors   p. 192
Size Sparrow; length 13 cm (5 in.).
Field Characters *Above*, crown and sides of neck black; a white nuchal
patch and glistening white cheeks. Back yellowish olive; rump grey.
Wings bluish with two white wing-bars. Tail bluish: tips of outer

rectrices, and outer web of the outermost, white. *Below*, throat and a broad band down middle of belly black; rest yellow (subspecies *monticolus*). Sexes alike. Very similar in pattern to Grey Tit (457): distinguished by olive back (*v.* grey), yellow underparts (*v.* ashy) and double wing-bar (*v.* single).

Status, Habitat, etc. Resident, locally common, subject to vertical movements. Breeds between 1200 and 3600 m; winters at lower levels generally, and down to the duars. All types of wooded country—light forest, orchards, hill-station gardens, etc. Pairs or small parties, often associated with other small birds. Lively and restless. Habits, behaviour and food as in Grey Tit. Call: also similar, perhaps louder and clearer. Song, a pleasant whistling *whichy-whichy* ... repeated 4 to 6 times with variations.

---

459 **COAL TIT** *Parus ater* Linnaeus   p. 208
Size Sparrow − ; length 10 cm (4 in.).
Field Characters A small crested tit, bluish grey above, pale rusty buff below. Head, throat and crest black; cheeks and a nuchal patch white; two white wing-bars (rows of spots) (subspecies *aemodius*). Sexes alike.
Status, Habitat, etc. Resident, common, subject to vertical movements. Breeds between 2500 and at least 3600 m; recorded in winter down to 1800 m. Parties often hunt in typical acrobatic fashion with itinerant flocks of small insectivorous birds, in spruce and deodar forest. Food: insects. Call: an incessant thin cheeping *tsi, tsi* while foraging.

---

460 **BLACK TIT** *Parus rubidiventris* Blyth   p. 192
Size Sparrow ± ; length 13 cm (5 in.).
Field Characters Distinguished from very similar Coal Tit by absence of double wing-bar, presence of a rufous patch on flanks, and slaty (not rufous-buff) underparts (subspecies *beavani*). Sexes alike.
Status, Habitat, etc. Resident, common, subject to vertical movements. Between 2700 and 4200 m in summer; down to 2200 m in winter. Fir, pine and juniper forest, and rhododendron scrub above timber-line (summer); oak forest (winter). One of the commonest high-elevation tits; often in mixed roving flocks of tinies. Forages mostly among the tree tops in typical tit manner. Food: insects, seeds. Call: a loud cheery *gypsie-bee* ... and some twittering and reeling notes. Song, a musical double whistle *whi-whee* ... normally repeated four times.

---

461 **BROWN CRESTED TIT** *Parus dichrous* Hodgson   p. 208
Size Sparrow − ; length 12 cm (5 in.).
Field Characters A plain grey-and-buff crested tit. *Above*, including crown

and pointed erect crest, brownish grey. A conspicuous whitish collar, interrupted on back. *Below*, throat greyish fulvous, rest brownish buff (subspecies *dichrous*). Sexes alike. May be casually mistaken for Yellow-naped Yuhina (304) q.v.

Status, Habitat, etc. Resident, fairly common, from 2700 up to timber-line in summer, down to 2200 m in winter. Mixed forest of oak, rhododendron, birch, fir, etc. Sociable, as other tits; a pair or so commonly in the mixed foraging flocks. Food: chiefly insects. Call: a characteristic thin high-pitched *zai* and some Goldcrest-like contact-notes.

462 BLACKSPOTTED YELLOW TIT *Parus spilonotus* Bonaparte

p. 192

Size Sparrow; length 14 cm (5½ in.).

Field Characters A sprightly black-and-yellow tit with yellow forehead and lores and pointed upstanding black crest. Long yellow supercilia meeting yellow nape-patch. Back olive-green streaked with black. *Below*, bright yellow with a broad black median band from chin to vent (subspecies *spilonotus*). Sexes practically alike.

Status, Habitat, etc. Resident, subject to vertical movements: scarce and local. Between 1600 and 2400 m (summer); down to 1400 or slightly lower (winter): light mixed forest and wooded cultivation environs. Ecology, behaviour and food as typical of tits. Call: song, a loud, spirited whistling *did-he-do-it did-he-do-it no-he-didn't* given from tree tops.

463 YELLOWBROWED TIT *Sylviparus modestus* Burton   p. 208

Size Sparrow − ; length 9 cm (3½ in.).

Field Characters Easily mistakable for a leaf warbler. *Above*, olive-green with a pale eye-ring; a short crest not visible unless erected. A short bright yellow supercilium more distinct when crest erected. *Below*, olive-buff (subspecies *modestus*). Sexes alike.

Status, Habitat, etc. Resident, fairly common, subject to vertical movements. Between 2100 and 4000 m in summer; down to 1500 m in winter. Fairly open conifer and broad-leaved forest, and dwarf scrub near timber-line (summer); heavy rhododendron and evergreen bushy jungle on hillsides (winter). A pair or so usually among itinerant foraging flocks of tinies. Acrobatic feeding postures typically tit-like; restless flitting in foliage with nervous wing-flicks deceptively like leaf warbler. Food: chiefly insects. Call: a feeble high-pitched *psit* or *tzee, tzee*. Song, an incessant thin shrill *zee-zi zee-zi zee-zi* . . .

464 FIRECAPPED TIT *Cephalopyrus flammiceps* (Burton) p. 96
Size Sparrow − ; length 9 cm (3½ in.).
Field Characters A tiny short-tailed yellowish olive bird reminiscent of a flowerpecker, with bright orange-scarlet forecrown. Wings brown, with two yellow bars. *Below*, chin and throat bright orange-scarlet paling to greenish saffron on breast and to pale yellow on belly. Female has yellow forehead (absent in winter), yellowish rump, and single wing-bar. Underparts pale yellowish olive and buff (subspecies *olivaceus*). Casually confusable with Yellowbrowed Tit (463).
Status, Habitat, etc. Little known; apparently rare and local. Specimens taken in winter in Jalpaiguri and Buxa duars, and in Sikkim and Bhutan (between 300 and 2300 m elevation) in open forest. Behaviour and actions like both leaf warblers and tits. Food: insects, leaf- and flower-buds. Call: song, a faint twittering reminiscent of White-eye's jingle, but more sustained.

465 REDHEADED TIT *Aegithalos concinnus* Gould p. 192
Size Sparrow − ; length 10 cm (4 in.).
Field Characters *Above*, crown chestnut: supercilium white; a broad black band through eye and ear-coverts. Back slaty grey; wings brown with a darker shoulder-patch. Tail brown, edged and tipped with white. *Below*, chin and sides of throat white, centre of throat ('bib') black; rest rusty buff (subspecies *rubricapilla*). Sexes alike. Young have white throat.
Status, Habitat, etc. Resident, common, subject to vertical movements. Breeds between 1400 and 3000 m; descends to 600 m in the foothills in winter: light deciduous forest and secondary growth of brambles etc.; sometimes deodar or pine forest. Very sociable, restless and fussy. Invariably in flocks, and commonly among the roving mixed hunting parties. Behaviour typically tit-like. Food: insects, berries. Call: a continual soft *trr-trr-trr* while foraging. Alarm, a rustling *prrri-prrri* taken up by all members of a party.

466 RUFOUSFRONTED TIT *Aegithalos iouschistos* (Hodgson)
p. 208
Size Sparrow − ; length 10 cm (4 in.).
Field Characters A tiny longish-tailed tit with a roundish silvery throat-patch, buffish 'centre parting' on crown, and ferruginous underparts. *Above*, forehead and a broad medial coronal band rusty buff; a broad black eye-stripe; sides of neck rusty buff; back grey. Wing with a dark shoulder-patch; tail with narrow white outer edge. *Below*, chin blackish;

centre of throat silvery white; malar stripe, sides of throat and rest of underparts rufous (subspecies *iouschistos*). Sexes alike.

Status, Habitat, etc. Resident, locally common, subject to erratic movements. Between 2700 and 3600 m at all seasons, some descending to 2400 m during cold spells. Low shrubby undergrowth of rose, barberry, bamboo, etc. in conifer and mixed forest. Behaviour and food very similar to Redheaded Tit (465). Call: a short sharp note as it moves about.

NUTHATCHES, WALL CREEPERS: Sittidae

467 **CHESTNUTBELLIED NUTHATCH** *Sitta castanea* Lesson

p. 209

Size Sparrow − ; length 12 cm (5 in.).

Field Characters Male. *Above*, bluish slaty; a black band from lores through eye to upper back; a prominent white malar patch. Expanded tail shows black rectrices with white subterminal spots. *Below*, chin white, rest dark chestnut-brown; under tail-coverts white and chestnut. Female similar, but underparts paler, more cinnamon (subspecies *cinnamoventris*).

Status, Habitat, etc. Resident, common. Duars and up to 1600 m— chiefly deciduous forest. Pairs or family parties, usually among the mixed hunting flocks. Searches crevices in bark for lurking insects, creeping up or down tree-trunks and spiralling round branches with mouse-like agility. Actions reminiscent of both woodpecker and tit. Food: insects, seeds, nuts. Call: a continual quiet *ti-ti-ti* while foraging. Song, a single loud clear whistle *chwhee* repeated unhurriedly.

468 **WHITECHEEKED NUTHATCH** *Sitta leucopsis* Gould

Size Sparrow − ; length 12 cm (5 in.).

Field Characters *Above*, crown and sides of nape black; face and broad supercilia white; rest dark bluish grey. Tail black with a white subterminal spot on the three outer pairs of rectrices. *Below*, rufous; lower flanks and vent chestnut (subspecies *przewalskii*). Sexes alike.

Status, Habitat, etc. Occurs in SE. Tibet and may extend into adjacent Arunachal Pradesh. Conifer forest.

469 **WHITETAILED NUTHATCH**

Sitta himalayensis Jardine & Selby   p. 209

Size Sparrow − ; length 12 cm (5 in.).

Field Characters Distinguished from all similar nuthatches by white patch at base of tail. *Above*, bluish slate. A black eye-stripe from lores to nape. Tail black with a prominent white patch at base; when expanded it shows white spots on the feathers. *Below*, throat buff shading to ochraceous on breast and deep rufous on lower parts (subspecies *himalayensis*). Sexes alike.

Status, Habitat, etc. Resident, common, subject to vertical movements. Breeds between 1500 and 3300 m; recorded in winter as low as 950 m: deciduous or evergreen broad-leaved forest, preferably mossy. Pairs or scattered parties commonly in the mixed foraging bird associations. Creeps jerkily along moss-covered branches or up and down around tree-trunks like a woodpecker, searching for lurking insects. Food: mainly insects; also nuts and seeds. Call: contact, a feeble mousy *chip-chip*. Song, spirited clear tit-like whistling *weet-wit-wit-wit-wit* ... quickly repeated several times.

470 **BEAUTIFUL NUTHATCH** *Sitta formosa* Blyth   p. 65

Size Sparrow ± ; length 15 cm (6 in.).

Field Characters A large, showy black-and-blue nuthatch. *Above*, brilliant blue streaked with lilac: back and rump pale blue. Wings black-and-blue with two white wing-bars and white edges to tertials. Tail blue: when expanded shows black rectrices with white subterminal spots. *Below*, sides of head buffish white with dark ear-coverts. Chin and upper throat creamy white darkening to rufous on belly and vent. Sexes alike.

Status, Habitat, etc. Resident, rare and local, subject to vertical movements. Recorded in summer between 1500 and 2100 m; in winter between 2000 and 330 m: deep forest. Pairs or small parties. Behaviour, food, etc. as of other nuthatches; specifically little known. Call: of typical pattern, somewhat lower pitched.

471 **VELVETFRONTED NUTHATCH** *Sitta frontalis* Swainson

p. 209

Size Sparrow − ; length 10 cm (4 in.).

Field Characters A small purplish blue nuthatch with velvety black forehead and supercilium, coral-red bill, orange-yellow orbital skin, and yellow eye. Tail black showing greyish terminal band when expanded. *Below*, chin and throat white grading to brownish tinged with lavender on belly. Female lacks black supercilium (subspecies *frontalis*).

Status, Habitat, etc. Resident, locally common. From plains level and duars up to 2000 m: broad-leaved forest and mixed bamboo jungle.

Plate 36, artist K. P. Jadav

GROSBEAKS, GREENFINCH, SUNBIRDS, WHITE-EYE, SNOW FINCH, SPARROWS, SPIDERHUNTER, MUNIA

1 R **SPOTTEDWINGED GROSBEAK,** *Mycerobas melanozanthos*
    Myna. Blackish rump (both sexes). ♀ Yellow supercilium. 600–3600 m.
          page 240

2 R **WHITEWINGED GROSBEAK,** *Mycerobas carnipes*   239
    Myna. Yellowish rump (both sexes). ♀ Breast grey. 1500–4000 m.

3 R **GREENFINCH,** *Carduelis spinoides*   240
    Sparrow – . 1500–3100 m.

4 R **INDIAN YELLOWBACKED SUNBIRD,** *Aethopyga siparaja*   234
    Sparrow – . ♂ Chin and throat scarlet. ♀ Outer rectrices tipped white. Up to 1800 m.

5 R **WHITE-EYE,** *Zosterops palpebrosa*   236
    Sparrow – . Up to 1500 m.

6 R **REDNECKED SNOW FINCH,** *Montifringilla ruficollis*   237
    Sparrow ± . Hindcrown, ear-coverts, sides of neck rufous. 4200–4800 m.

7 R **NEPAL YELLOWBACKED SUNBIRD,** *Aethopyga nipalensis*   233
    Sparrow – . ♂ Throat and central rectrices metallic blue-green. ♀ As 4, with white-tipped outer rectrices. 300–3600 m.

8 R **CINNAMON TREE SPARROW,** *Passer rutilans*   237
    Sparrow. ♀ Dark-streaked above. Whitish supercilium. Double wing-bar. Up to 4000 m.

9 R **TREE SPARROW,** *Passer montanus*   236
    Sparrow. ♂♀ Alike. Crown chocolate. A black spot on sides of head. Up to 3000 m.

10 R **STREAKED SPIDERHUNTER,** *Arachnothera magna*   235
    Sparrow. Streaked plumage. Long curved black bill. 300–1500 m.

11 R **SPOTTED MUNIA,** *Lonchura punctulata*   238
    Sparrow – . Up to 2300 m.

JPIrani 1973

Plate 37, artist J. P. Irani

ROSEFINCHES, BUNTINGS, BULLFINCHES, CROSSBILL

1 R **BEAUTIFUL ROSEFINCH,** *Carpodacus pulcherrimus*   page 244
Sparrow. ♂ Crown broadly black-streaked. ♀ Fulvous-brown above; fulvous-white below, streaked blackish. 2000–4500 m.

2 M **LITTLE BUNTING,** *Emberiza pusilla*   250
Sparrow – . Black loop round chestnut ear-coverts. Up to 1800 m.

3 R **REDHEADED ROSEFINCH,** *Propyrrhula subhimachala*   246
Bulbul. Heavy bill. 1800–4200 m.

4 M **CHESTNUT BUNTING,** *Emberiza rutila*   248
Sparrow ± . Duars

5 R **REDBREASTED ROSEFINCH,** *Carpodacus puniceus*   245
Bulbul ± . ♀ Dark-streaked grey-brown above; brown-streaked buff below. Rump yellow. 2700–5200 m.

6 R **BEAVAN'S BULLFINCH,** *Pyrrhula erythaca*   248
Sparrow + . Crown grey. 2000–3800 m.

7 M **ROSEFINCH,** *Carpodacus erythrinus*   242
Sparrow. Belly and vent pale (both sexes). Up to 4000 m.

8 R **CROSSBILL,** *Loxia curvirostra*   246
Sparrow. 1500–4000 m.

9 R **BROWN BULLFINCH,** *Pyrrhula nipalensis*   247
Sparrow + . Crown scaly-patterned brown. 1500–4000 m.

10 R **PINKBROWED ROSEFINCH,** *Carpodacus rhodochrous*   244
Sparrow. ♀ Prominent pale supercilium. 1200–4000 m.

11 R **CRESTED BUNTING,** *Melophus lathami*   251
Sparrow. ♀ Edge of wing and outer rectrices rufous. Up to 1800 m.

12 R **BLANFORD'S ROSEFINCH,** *Carpodacus rubescens*   243
Sparrow. ♂ Double crimson wing-bar. Vent grey. ♀ Brown above with reddish rump. Faint double wing-bar. Unstreaked greyish underparts; white-edged under tail-coverts. 1300–3800 m.

Ecology largely as of other nuthatches. Call: a rapid series of loud, high-pitched cheeping whistles.

472 WALL CREEPER *Tichodroma muraria* (Linnaeus) p. 65

Size Sparrow; length 17 cm (6½ in.).

Field Characters A sober grey nuthatch-like bird easily identified in flight by the brilliant white-spotted crimson wings which are continually flicked open and shut while creeping up vertical cliffs. Throat and upper breast black in summer, white in winter (subspecies *nipalensis*). Sexes almost alike.

Status, Habitat, etc. Resident, fairly common but local and sporadic, subject to altitudinal movements. Breeds mostly above 3300 m and up to permanent snow-line; descends lower and into the adjacent plains in winter: rocky gorges and vertical cliffs. Solitary or pairs. Clings to cliffs and climbs in jerky zigzagging spurts, searching the unevennesses and under moss and lichens for prey. Flight hoopoe-like, with alternate flaps and pauses: also reminiscent of butterfly. Food: chiefly insects and spiders. Call: a soft plaintive cheeping. Song, a thin slow ascending phrase of four sliding notes, the last very thin.

TREE CREEPERS: Certhiidae

473 MANDELLI'S TREE CREEPER *Certhia familiaris* Linnaeus

p. 65

Size Sparrow − ; length 12 cm (5 in.).

Field Characters *Above*, fulvous brown spotted with rufous on crown and back; rump ferruginous; a white supercilium. A short wing-bar and oblique buff band across wing. Tail brown, longish, pointed, unbarred. *Below* white, lower flanks tinged brown (subspecies *mandellii*). Sexes alike.

Distinguished from *C. discolor* (474) by white throat and breast (*v.* brown); from *C. nipalensis* (475) by white breast and earth-brown flanks (*v.* buff breast, ferruginous flanks).

Status, Habitat, etc. Resident, common, subject to vertical movements. Breeds from 2700 up to timber-line (*c.* 4000 m); winters between 3000 and 1700 m: conifer and broad-leaved forests. Solitary or pairs, commonly among the mixed foraging flocks. Creeps up tree-trunks in short jerky woodpecker-like spurts, searching the bark crevices for prey. Food insects, spiders. Call: a thin *tsee*; song, a very thin short descending phrase.

474 SIKKIM TREE CREEPER *Certhia discolor* Blyth   p. 209
Size Sparrow — ; length 12 cm (5 in.).
Field Characters *Above*, dark brown streaked with fulvous; supercilium
fulvous. A short bar and an oblique buff band across wing. Rump
ferruginous; tail rufous-brown. *Below*, throat and breast tawny brown,
belly paler, vent ochraceous (subspecies *discolor*). Sexes alike. Easily
identified by dark throat (*v.* white in other creepers).
Status, Habitat, etc. Resident, fairly common, subject to vertical
movements. Recorded in summer between 1800 to 2700 m; in winter
from 3600 m down to 700 m: mossy oak and rhododendron forest,
also conifers. Ecology, behaviour and food as of other tree creepers.
Call: a sharp, quick-repeated *chip-chip-chip-chip*. Song, thin, short,
typical of the genus.

475 NEPAL TREE CREEPER *Certhia nipalensis* Blyth
Size Sparrow — ; length 12 cm (5 in.).
Field Characters *Above*, crown and upper back blackish streaked with
fulvous; a very broad buff supercilium. Mantle rusty brown with
blackish scale-like markings; rump rusty. Wings with a buff bar and a
buff oblique band. Tail brown, unbarred. *Below*, throat whitish, breast
and belly cream-buff; flanks tawny-olive, rusty posteriorly. Sexes alike.
Distinguished from *discolor* (474) by whitish throat; from *familiaris*
(473) by buff breast and tawny flanks (*v.* white and greyish).
Status, Habitat, etc. Resident, uncommon, subject to vertical movements. Recorded in summer between 2700 and 3500 m; in winter between
3500 and 1500 m: mixed broad-leaved forest and conifers. Ecology,
behaviour and food as of the genus. Call: unrecorded.

PIPITS, WAGTAILS: Motacillidae

476 INDIAN TREE PIPIT *Anthus hodgsoni* Richmond   p. 209
Size Sparrow ± ; length 15 cm (6 in.).
Field Characters *Above*, greenish brown streaked with darker brown;
supercilium, double wing-bar and outer rectrices whitish. *Below*,
buffish white, boldly streaked with dark brown on breast and flanks
(subspecies *hodgsoni*). Sexes alike. Not easily distinguished from
European Tree Pipit (*A. trivialis*) which has browner upperparts and
bolder streaking on breast.
Status, Habitat, etc. Common summer visitor. Breeds up to 4000 m;
winters from about 2000 m down through the duars and Indian plains.
Rocky ground and grass- and bracken-covered hillsides, and open

forest glades (summer); wooded country and groves (winter). Pairs
or loose flocks. Runs on ground in search of food, slowly wagging
tail; flies up into trees on disturbance. Flight jerky and undulating as
typical of pipits. Food: insects, grass seeds. Call: a single *tseep*.
Song, lark-like, harsher and more wheezy; uttered on wing in display
flight.

477 **BLYTH'S PIPIT** *Anthus godlewskii* (Taczanowski)   p. 209
Size Sparrow ± ; length 15 cm (6 in.).
Field Characters *Above*, dark brown marked with fulvous. Tail dark
brown with white outer rectrices conspicuous in flight and when alighting. *Below*, buff streaked with brown on breast. Sexes alike.
      Indistinguishable from Paddyfield Pipit (*A. novaeseelandiae*) in the
field except by voice and calls.
Status, Habitat, etc. Status uncertain: presumably an autumn passage
migrant from eastern Asia to the Indian Peninsula. Recorded in Sikkim
in September and October between 2100 and 4500 m on swampy land.
Call: said to be peculiarly harsh and very different from other pipits',
and diagnostic.

478 **VINACEOUSBREASTED PIPIT** *Anthus roseatus* Blyth
Size Sparrow ± ; length 15 cm (6 in.).
Field Characters Summer: *Above* grey with blackish streaks; a distinct
vinous-buff supercilium. Wings brown with greenish edges and two
pale bars. Tail brown with whitish outer edges. *Below*, throat and
breast pale pink or vinaceous-buff, faintly streaked on breast. Rest
buffish, streaked on flanks with dark brown. Sexes alike. Autumn:
Above, olive-brown instead of grey; supercilium often tinted yellow.
Below, pink of throat fainter; breast heavily streaked. Field identification difficult.
Status, Habitat, etc. Common summer visitor. Breeds mostly above
timber-line, above 4200 m: alpine meadows and boulder-strewn
slopes, especially marshy ground. Winters from 1500 m down through
the foothills, duars and plains, near marshes and jheels. Habits etc.
typical of the genus. Call: a sharp single or double *tseep* on flushing.
Song, a series of pleasant fading *tsuli-tsuli-tsuli* ... usually given in
display flight: said to be reminiscent of Redwinged Bush Lark's
(*Mirafra erythroptera*).

479 **FOREST WAGTAIL** *Motacilla indica* Gmelin   p. 209
Size Sparrow; length 17 cm (6½ in.).
Field Characters *Above*, olive-brown. A pale supercilium. Wings blackish

with two prominent yellowish bands across coverts. Tail blackish with the outer feathers white. *Below*, pale yellowish white; blackish collar across lower throat; a second interrupted band on breast. Sexes alike.

Pipit-like appearance, double black gorget on breast and yellowish double wing-bar diagnostic.

Status, Habitat, etc. Winter straggler or scarce passage migrant; chiefly the duars, in forest glades and clearings, and mixed bamboo jungle. More arboreal than other wagtails. Normally solos. Forages quietly on ground, swaying tail and hind part of body laterally. Flies up into overhanging branch on disturbance, wagging tail slowly up and down and from side to side. **Food:** insects. **Call:** in winter a characteristic finch-like *pink* or *pink-pink*.

---

480 **GREY WAGTAIL** *Motacilla caspica* (Gmelin)　　　　　p. 209

Size Sparrow, with long tail; length 17 cm (6½ in.).

Field Characters Male (summer). *Above*, head and back grey; rump greenish yellow; a white supercilium. Wings blackish with white edges to tertials showing as a prominent V on back at rest. Tail blackish with white outer edges. *Below*, throat black with white malar stripes; rest bright yellow. In winter throat buffish, underparts paler yellow (brighter on vent), and white V on back indistinct. Female (summer) has buff throat mottled with black, and paler yellow underparts (subspecies *caspica*). Sexes indistinguishable in winter.

Status, Habitat, etc. Summer (breeding) visitor to the western Himalayas. Only sparse winter visitor east of Nepal, chiefly duars and foothills, rarely up to 2000 m: neighbourhood of water—rocky streams and trickles, rock pools in river-beds, forest streams, etc. Solos or widely separated pairs. Runs about to feed on ground, wagging its tail incessantly. Sometimes captures midges etc. by agile aerial sorties from a rock amid stream. **Food:** chiefly insects. **Call:** a sharp *chi-cheep*, *chi-cheep* uttered on wing. Song, a thin rapid twitter *chi-chi-chi-chi-chi-chi* given in display flight.

---

481 **HODGSON'S PIED WAGTAIL**
　　Motacilla alba alboides Hodgson　　　　　p. 209

Size Bulbul; length 18 cm (7 in.).

Field Characters *Above*, forehead and round eyes white; crown, nape, ear-coverts, all round neck, entire back and rump black. Wings black with large white patches on coverts and tertials. Tail black with white outer rectrices. *Below*, chin, throat and breast black; rest white. Sexes almost alike.

Status, Habitat, etc. Summer (breeding) visitor and vertical and short-range migrant. Breeds between 2700 and 5000 m; winters mostly below 1500 m. Affects river-beds, rocky streams and wet fields. Habits, behaviour and food typical of the genus. Call: a sharp *chicheep* usually uttered in flight.

482 STREAKEYED PIED WAGTAIL
Motacilla alba ocularis Swinhoe

Size Bulbul; length 18 cm (7 in.).

Field Characters Male (winter). *Above*, forehead and face white with a diagnostic black streak through eye to nape. Hindcrown and nape black; entire back and rump ashy grey. Wings blackish; coverts and tertials broadly white-margined. Tail blackish with white outer rectrices. *Below*, white with a large crescent-shaped black patch on breast. Sides of breast, and flanks, ashy. Sexes almost alike. In summer neck (all round) and throat also black.

Status, Habitat, etc. Uncommon winter visitor. Duars and foothills—open country near water, rivers, tanks, paddy fields, compounds, etc. Habits and food typical of the genus. Call: as in 481.

483 SWINHOE'S PIED WAGTAIL
Motacilla alba baicalensis Swinhoe

Size Bulbul; length 18 cm (7 in.).

Field Characters Like Streakeyed Pied (482) but chin, throat and upper breast white at all seasons, uniformly with rest of underparts.

Status, Habitat, etc. Uncommon winter visitor to the duars. Habits, food and calls as in 481.

484 LARGE PIED WAGTAIL *Motacilla maderaspatensis* Gmelin

p. 209

Size Bulbul; length 21 cm (8½ in.).

Field Characters A large black-and-white wagtail of nearly the colour pattern of Magpie-Robin, with a conspicuous white supercilium from lores to nape. Differs from Hodgson's Pied (481) by larger size and in having the black of crown projecting in a point over forehead to base of bill. (In 481 the forehead is broadly white.) Also has the white wing-patch larger. Female has the black portions duller and browner.

Status, Habitat, etc. Resident. Duars and foothills up to 900 m: smooth-running rocky streams, bunded irrigation tanks, masonry wells and neighbourhood of cultivation. Usually pairs. Typical wagtail habits and behaviour. Food: insects. Call: a loud *chiz-zit* chiefly in flight. Song, a clear high-pitched jumble of loud, pleasant whistling notes reminiscent of Magpie-Robin's song.

FLOWERPECKERS: Dicaeidae

485 YELLOWVENTED FLOWERPECKER
Dicaeum chrysorrheum Temminck

Size Sparrow − ; length 9 cm (3½ in.).

Field Characters An olive-green flowerpecker with buffish white under-parts distinctly streaked with dark brown, and bright yellow vent. A dark moustachial streak (subspecies *chrysochlore*). Sexes alike. Streaked underparts diagnostic.

Status, Habitat, etc. Resident; less common than south of Brahmaputra river. Duars, foothills and up to 2000 m: open jungle, forest edges, orchards, etc. Solitary or small parties. Frequents canopy foliage of trees, especially with clumps of mistletoe plant parasites. Food: berries, chiefly of mistletoe; also flower nectar, insects, spiders. Call: unrecorded.

486 YELLOWBELLIED FLOWERPECKER
Dicaeum melanoxanthum (Blyth)   p. 209

Size Sparrow − ; length 12 cm (5 in.).

Field Characters A relatively large black-and-yellow flowerpecker. Male. *Above*, slaty black; white spots in tail. *Below*, sides of head, neck and breast slaty black; a white band from chin down centre of throat and breast; rest bright yellow. Female like male but black replaced by olive-brown, paler on sides of head, neck and breast; central band greyish white; belly and vent yellow.

Status, Habitat, etc. Recorded in summer (breeds?) between 2700 and 3600 m; in winter between 1800 and 1600 m: tall trees in open forest and clearings. Habits little known. Keeps to the foliage canopy of tall trees. Has been observed making aerial fly-catching sorties. Food: probably as of other flowerpeckers, chiefly berries. Call: unrecorded.

487 PLAINCOLOURED FLOWERPECKER
Dicaeum concolor Jerdon

Size Sparrow − ; length 8 cm (3 in.).

Field Characters A tiny plain-coloured flowerpecker, brownish olive-green above, greyish buff below. Similar to the commoner Tickell's Flowerpecker of the Indian plains but with a blackish (*v.* pink) bill and pale lores and supercilium (subspecies *olivaceum*). Sexes alike.

Status, Habitat, etc. Resident, common. Duars and foothills between 700 and 1000 m, sparingly up to at least 1400 m: groves of trees, forest glades and edges, orchards, etc. Usually pairs; very active and restless, hopping amongst parasitic mistletoe clumps in tall trees. Food:

chiefly berries; also nectar, insects, spiders. Call: a sharp *chip*, *chip*, *chip* in flight; also a characteristic twittering.

488 SCARLETBACKED FLOWERPECKER
Dicaeum cruentatum (Linnaeus)   p. 209

Size Sparrow − ; length 7 cm (2½ in.).

Field Characters A distinctive brilliantly coloured black-and-crimson flowerpecker. **Male.** *Above*, crown, back and rump bright crimson; sides of head black. Wings and tail glossy blue-black. *Below*, black with a buff band down centre of throat, breast and belly. Flanks grey. **Female.** *Above*, olive; rump crimson. Tail black; wings dark brown. *Below*, buff with grey sides (subspecies *cruentatum*).

Status, Habitat, etc. Resident, common. Duars and foothills at fairly low elevations: open forest, orchards, groves in cultivation, etc. Habits, behaviour, food, etc. as of the genus. Call: a loud *tchik-tchik* recalling a Tailor Bird's; also the characteristic twittering while feeding.

489 FIREBREASTED FLOWERPECKER *Dicaeum ignipectus* (Blyth)
pp. 49, 145

Size Sparrow − ; length 7 cm (2½ in.).

Field Characters **Male.** *Above*, metallic greenish black. *Below*, rich buff, flanks olive; a scarlet patch on breast; a black patch on belly. **Female.** *Above*, olive-green, yellower on rump. *Below*, pale buff washed with olive on sides (subspecies *ignipectus*).

Status, Habitat, etc. Resident, common, subject to vertical movements. Breeds between 1400 and 3000 m; winters from 2000 m down through the foothills to 750 m or so: tall forest as well as secondary growth, orchards, groves, etc. Solos or pairs, almost invariably on *Loranthus* clumps parasitizing trees. Food: insects, spiders, nectar and berries especially of the mistletoe family. Call: a single *chip* in flight; a shrill twitter *titty-titty-titty*.

SUNBIRDS: Nectariniidae

490 RUBYCHEEK *Anthreptes singalensis* (Gmelin)   p. 209

Size Sparrow − ; length 10 cm (4 in.).

Field Characters **Male.** *Above*, brilliant metallic green; ear-coverts copper-coloured bordered below by a metallic violet-purple malar stripe. *Below*, throat and breast rufous; rest lemon-yellow. **Female.** *Above*, olive-green, yellowish on wing. *Below*, like male (subspecies *rubinigentis*).

Status, Habitat, etc. Resident, fairly common. From plains level through the duars and foothills up to 700 m: scrub jungle, secondary evergreen forest, and forest glades. Solos, sometimes small parties, moving in the tree tops and foraging energetically like tits or white-eyes. Food: insects, flower nectar. Call: a shrill, loud chirp in flight.

491 MRS GOULD'S SUNBIRD *Aethopyga gouldiae* (Vigors)

pp. 49, 145

Size Sparrow − ; length ♀ 10 cm (4 in.); ♂ 15 cm (6 in.), including long pointed tail.

Field Characters Male. *Above*, crown, ear-coverts and a patch on each side of neck metallic purple-blue; sides of head, supercilium, neck, nape and back crimson; rump bright yellow. Tail (long, graduated, pointed) metallic purple-blue. *Below*, throat metallic purple-blue; rest yellow, streaked with scarlet on breast; vent olivaceous. Female. *Above*, head and nape grey; rest olive, yellowish on rump. Tail not elongated. *Below*, throat pale grey; rest yellow (subspecies *gouldiae*).

Status, Habitat, etc. Resident, locally common, subject to vertical movements. Breeds between 1800 and 3300 m; withdraws lower in winter, reaching the duars and base of the hills: oak and conifer forest. Solos or pairs. Active and vivacious, visiting flowers in quest of nectar and flitting ceaselessly from shrub to shrub. Food: nectar, insects, spiders. Call: a scissors-like *tzit-tzit*.

492 NEPAL YELLOWBACKED SUNBIRD
Aethopyga nipalensis (Hodgson)   p. 224

Size Sparrow − ; length ♀ 10 cm (4 in.); ♂ 15 cm (6 in.) including long pointed tail.

Field Characters Male. *Above*, crown and nape metallic blue-green; sides of neck and upper back maroon. Middle back and wings olive-green; rump bright yellow. Tail (pointed, graduated) metallic blue-green. *Below*, cheeks black; throat metallic blue-green; rest bright yellow, streaked with scarlet on breast. Female olive-green. Tail not elongated, outer rectrices white-tipped (subspecies *nipalensis* and *koelzi*).

Status, Habitat, etc. Resident, common, subject to vertical movements. Breeds between 1800 and 2700 m; recorded in winter as high as 3600 and down to 300 m: heavy oak and rhododendron forest, and scrub jungle. Habits, behaviour, etc. as of the genus. Food: mainly flower nectar. Call: a sharp *dzit*.

493 BLACKBREASTED SUNBIRD *Aethopyga saturata* (Hodgson)
pp. 49, 145

Size Sparrow − ; length ♀ 10 cm (4 in.); ♂ 15 cm (6 in.) including long pointed tail.

Field Characters Male. *Above*, crown and nape metallic purple; back and sides of neck maroon. Wings blackish. A narrow yellow rump-band; tail-coverts and tail (pointed, graduated) metallic purple. *Below*, throat and breast dull blackish; a broad metallic purple malar streak; rest greyish olive. Female, a nondescript olive-green without elongated tail, difficult to identify when unaccompanied by male (subspecies *saturata*).

Status, Habitat, etc. Resident, common, subject to vertical movements. Up to 2000 m in summer; down to 450 m in winter. Chiefly evergreen jungle and forest outskirts. Habits, behaviour and food as of the genus. Call: not specifically described.

494 INDIAN YELLOWBACKED SUNBIRD
Aethopyga siparaja (Raffles)   p. 224

Size Sparrow − ; length ♀ 10 cm (4 in.); ♂ 15 cm (6 in.) including long pointed tail.

Field Characters Male. *Above*, crown metallic green; sides of neck, and back, dark crimson; rump bright yellow. Tail (pointed, graduated) metallic green. *Below*, chin and throat scarlet with metallic purple malar streaks; belly yellowish olive. Female entirely olive, more yellowish below. Tail not elongated (subspecies *seheriae* and *labecula*).

Status, Habitat, etc. Resident, common, subject to vertical movements. Breeds from the foothills up to 1800 m; more common in winter in the duars and adjacent plains. Dense evergreen forest and secondary jungle. Habits, food, etc. as of the genus. Call: very like noise of the rapid opening and shutting of scissor blades. Song, a chirping trill.

495 FIRETAILED YELLOWBACKED SUNBIRD
Aethopyga ignicauda (Hodgson)   pp. 49, 145

Size Sparrow − ; length ♀ 10 cm (4 in.); ♂ 15 cm (6 in.) including long pointed tail.

Field Characters Male. *Above*, crown metallic purple; sides of head, nape, back, tail-coverts and tail bright red. Rump bright yellow. *Below*, throat metallic purple; rest yellow, washed with scarlet on breast. Bright red elongated pointed tail diagnostic. Female olive; more yellow on rump and belly. Tail not elongated (subspecies *ignicauda*).

Status, Habitat, etc. Resident, common, subject to vertical movements. Breeds between 3000 and 4000 m; winters from 2900 m down to 1200 or so. Affects open conifer forest with dense undergrowth of

rhododendron, juniper and barberry bushes. Habits and food as of the
genus. Call: song, a high-pitched monotonous *dzidzidzidzidzidzi* ...
constantly repeated.

496 LITTLE SPIDERHUNTER *Arachnothera longirostris* (Latham)

Size Sparrow − ; length (including long bill) 14 cm (5½ in.).
Field Characters Reminiscent of a large olive-and-yellow female sunbird,
with a very long curved bill. *Above*, olive. Tail dark brown, tipped with
white. *Below*, throat and breast greyish white; belly yellow, with orange
tufts on flanks (subspecies *longirostris*). Sexes alike.
Status, Habitat, etc. Resident, locally common. From plains level
through the duars and lower foothills up to 1500 m. Evergreen biotope:
dense forest, and secondary growth interspersed with wild banana
patches. Habits similar to sunbirds'. Very fond of nectar and an
important cross-pollinator for wild banana and other adapted flowers.
Food: flower nectar, insects, spiders. Call: a harsh rather metallic
chee-chee. Song, a metallic *which-which-which-which* ... repeated for
a couple of minutes at a time.

497 STREAKED SPIDERHUNTER *Arachnothera magna* (Hodgson)

p. 224

Size Sparrow; length (including long bill) 17 cm (6½ in.).
Field Characters A large boldly streaked yellowish olive sunbird with very
long and stoutish curved black bill, and short yellow legs. *Above*,
yellowish olive streaked with black. Tail tipped with buffish spots,
with a blackish subterminal band. *Below*, yellowish white with bold
black shaft-streaks (subspecies *magna*). Sexes alike.
Status, Habitat, etc. Resident, locally common, subject to vertical
movements. Submontane plains, duars and foothills, normally between
300 and 1500 m. Evergreen biotope: dense forest, abandoned overgrown
cultivation clearings, and patches of wild banana plants. Habits similar
to sunbirds'. Often hovers while probing flowers for nectar or to take
spiders from their web. Food: flower nectar, insects, spiders. Call:
a sharp, metallic *chirik*, *chirik* while feeding and on the wing. Song,
rather soft, rapid and monotonous, commencing slowly and clearly
and gaining speed.

WHITE-EYES: Zosteropidae

498 WHITE-EYE *Zosterops palpebrosa* (Temminck)   p. 224
Size Sparrow – ; length 10 cm (4 in.).
Field Characters *Above*, olive-yellow; a very distinct white eye-ring (whence sometimes called 'Spectacle Bird'). Lores and a patch under eye blackish. *Below*, throat and under tail-coverts bright yellow; rest greyish white (subspecies *palpebrosa*). Sexes alike.
Status, Habitat, etc.   Resident, common, subject to seasonal movements. Plains, duars and foothills, up to 1500 m or slightly higher: forest, village groves, orchards, gardens, etc. Arboreal. Pairs, parties, sometimes considerable flocks, associated with other small birds. Clings upside down and in all positions to sprigs and flowers in quest of insect prey and nectar. Food: insects, nectar, flower buds, berries, pulpy fruit, etc. Call: a feeble, plaintive *cheer*, constantly uttered. Song, a tinkling jingle commencing softly, growing louder and soon fading out; reminiscent of Verditer Flycatcher's song.

WEAVER BIRDS: Ploceidae

499 HOUSE SPARROW *Passer domesticus* (Linnaeus)
Size Length 15 cm (6 in.).
Field Characters Familiar and well known. **Male.** Distinguished from Tree Sparrow (next) by grey crown (*v.* chocolate), chestnut nape and upper back, grey-brown rump, and black breast and throat. **Female.** *Above*, greyish brown, streaked with fulvous and dark brown on back. A pale supercilium. *Below*, plain fulvous white (subspecies *indicus*).
Status, Habitat, etc.   Resident but curiously patchy and uncommon at elevations above 1300 m where largely replaced by Tree Sparrow. As elsewhere, affects human habitations and cultivation. Often large flocks in winter. Familiar, cheeky, and completely at home in human environments. Food: grain, seeds, insects and kitchen scraps. Call: noisy chirruping; a loud monotonous *cheer, cheer, cheer* ... by male in breeding season.

500 TREE SPARROW *Passer montanus* (Linnaeus)   p. 224
Size Sparrow; length 15 cm (6 in.).
Field Characters *Above*, crown and nape chocolate-brown. Sides of head white, with a black patch on ear-coverts. Rest brown, streaked with black on back. *Below*, chin and centre of throat black; rest greyish white (subspecies *malaccensis*, *tibetanus* and *hepaticus*). Sexes alike.

Status, Habitat, etc. Resident, common, locally subject to vertical movements. From plains level through duars and foothills up to 3000 m. Affects human settlements, monasteries, dzongs and surrounding fields and cultivation. Takes the place of *Passer domesticus* as a house bird. Of the same familiar habits; perhaps less cheeky. Food: grain, seeds, insects, kitchen scraps. Call: chirping like House Sparrow's, somewhat more musical.

501 CINNAMON TREE SPARROW *Passer rutilans* Temminck

p. 224

Size Sparrow; length 15 cm (6 in.).

Field Characters Male. *Above*, bright rufous-chestnut, streaked with black on back. A broad and a narrow white wing-bar. *Below*, chin and centre of throat black, sides of throat pale yellow. Female. *Above*, brown, streaked with dark brown on back; a conspicuous whitish supercilium and double white wing-bar. *Below*, pale yellowish ashy (subspecies *cinnamomeus*).

Status, Habitat, etc. Resident, locally common, subject to vertical movements. Breeds between 1800 and 4000 m; winters through the foothills and duars into the adjacent plains. Affects light forest of oak, rhododendron, etc. and vicinity of hill villages, monasteries and dzongs, often taking the place of House Sparrow and sharing it with Tree Sparrow. Habits etc. of all three similar. Food: grain, seeds, insects, berries. Call: *chilp ... chilp ...* similar to but softer and pleasanter than House Sparrow's; also a thin *swee ... swee ...* recalling Indian Robin's.

502 REDNECKED SNOW FINCH *Montifringilla ruficollis* Blanford

p. 224

Size Sparrow ± ; length 15 cm (6 in.).

Field Characters *Above*, forehead and supercilium dingy white; a dark streak through eye. Hindcrown, ear-coverts and sides of neck rufous. Back pale brown with darker streaks. Wings brown with a prominent white shoulder-patch and largely white secondaries. Tail: central rectrices brown, outer grey, all with a broad, white subterminal band and blackish tips. *Below*, throat white, sides of throat rufous; a blackish malar streak. Rest creamy white. Sexes alike.

Status, Habitat, etc. Recorded in N. Sikkim in October and December at 4200–4800 m; also near Darjeeling in October. Affects open gravel plains and grassy plateaus. Small flocks in Tibetan steppe facies. Food: seeds, insects. Call: not recorded in winter.

503 **BLANFORD'S SNOW FINCH** *Montifringilla blanfordi* Hume
Size Sparrow ± ; length 15 cm (6 in.).
Field Characters Similar to Rednecked Snow Finch (502) q.v.;
distinguished from it by black markings on face, white ear-coverts
(*v.* rufous), unstreaked upperparts, lack of white shoulder-patch, and
black throat (*v.* white) (subspecies *blanfordi*). Sexes alike.
Status, Habitat, etc. Recorded in N. Sikkim in September, October and
December at 4200 m. Affects Tibetan steppe country and cultivation in
such. Flocks; sometimes large ones in winter. Food: seeds, insects.
Call: not recorded in winter.

504 **WHITEBACKED MUNIA** *Lonchura striata* (Linnaeus) p. 53
Size Sparrow − ; length 10 cm (4 in.).
Field Characters A small brown-and-white finch with heavy bluish conical
bill and pointed tail. *Above*, forehead, wings and tail blackish. Back
brown with fine pale shaft-streaks; rump white. *Below*, throat and
breast blackish; belly greyish white finely brown-streaked (subspecies
acuticauda). Sexes alike.
Status, Habitat, etc. Resident, uncommon, subject to seasonal movements. From the base of the hills and duars up to 1800 m locally: open
country, secondary growth and scrub jungle, often near cultivation.
Flocks of 8–15; sometimes large. Gleans on the ground and in harvested
fields. Roosts communally with weaver birds etc. among thickets and
sugarcane fields. Food: grass seeds, insects. Call: a quiet plaintive
cheeping and twitter.

505 **SPOTTED MUNIA** *Lonchura punctulata* (Linnaeus) p. 224
Size Sparrow − ; length 10 cm (4 in.).
Field Characters *Above*, chocolate-brown with pale shaft-streaks; rump
barred with white; tail-coverts and tail golden fulvous. *Below*, sides of
head, neck, chin, and throat chestnut; breast and flanks speckled
black-and-white; belly white (subspecies *punctulata*). Sexes alike.
Status, Habitat, etc. Resident, common. From the base of the hills and
duars up to 2300 m: open scrub country, grassland, gardens and cultivation. Flocks; sometimes of up to a hundred or more. Feeds by gleaning
on the ground or from growing grass stems. Flight undulating, in loose
disorderly rabbles. Roosts communally with weavers and other munias.
Food: grass seeds, insects. Call: a petulant-sounding *kitty-kitty-
kitty*; also a short husky whistle.

FINCHES: Fringillidae

506 ALLIED GROSBEAK *Mycerobas affinis* (Blyth)　　　p. 193
Size Myna; length 22 cm (8½ in.).
Field Characters A large black-and-yellow massive-billed finch. Male.
Above, head, wings and tail deep black; a yellow hind collar; centre of
back, and rump, yellow. *Below*, throat black; rest yellow. Female.
Above, head dark grey; rest olive-green, more yellowish on collar and
rump. Wings and tail blackish. *Below*, throat grey; rest yellowish olive.
Superficially similar to female *M. carnipes* (507) but lacks the white
wing-patch and has the grey restricted to throat.
Status, Habitat, etc. Resident, locally common, subject to vertical
movements. Breeds mostly between 3000 and 4000 m; descends lower
in winter, occasionally to 1800 m. Affects mixed conifer and broad-
leaved forest, sometimes wandering into dwarf rhododendron and
junipers above tree-limit. Pairs or small flocks. Feeds in trees as well
as on ground. Food: seeds, kernels, insects. Call: alarm, a double
note *kurr-kurr* like two stones struck together. Song, a loud musical
whistle of 5 to 7 notes.

507 WHITEWINGED GROSBEAK *Mycerobas carnipes* (Hodgson)
　　　　　　　　　　　　　　　　　　　　　　　　　　p. 224
Size Myna; length 22 cm (8½ in.).
Field Characters A large dark finch with massive bill and white wing-
patches, particularly conspicuous in flight. Male. *Above*, black with a
white wing-patch, yellowish olive rump and some spots of same colour
on secondaries. *Below*, throat and breast black; belly yellow.
Distinguished from male *M. melanozanthos* (508) by yellowish rump
(*v.* blackish) and more extensive black on underparts. Female. *Above*,
head and back grey; rump olive-yellow. Wings grey-brown with a white
patch, yellowish bar, and pale yellow and whitish pattern on secondaries.
Below, sides of head, throat and breast grey lightly mottled with buff.
Flanks and lower belly yellow. Distinguished from female *M. affinis*
by white wing-patch (subspecies *carnipes*).
Status, Habitat, etc. Resident, common, subject to vertical movements.
Breeds between 3000 and 4000 m; descends lower in winter, occasionally
down to 1500 m: dwarf juniper and fir forest (summer); mixed conifer
and broad-leaved forest at lower elevations (winter). Pairs or flocks,
noisily cracking hard juniper stones to feed on the kernels. Food:
seeds, kernels, berries, etc. Call: *croak-et-et* uttered from a prominent
perch.

508 SPOTTEDWINGED GROSBEAK
Mycerobas melanozanthos (Hodgson)   p. 224
Size Myna; length 22 cm (8½ in.).

Field Characters A large dumpy black-and-yellow finch with massive bill and prominent white spots on wings. Male similar to male *M. carnipes*; distinguished from it by blackish rump (*v.* yellowish olive), black on underparts confined to throat and breast, and much brighter yellow belly. Female. *Above*, blackish like male, but crown, nape and back streaked with yellow; a distinct yellow supercilium. *Below*, entirely yellow, boldly streaked with blackish.

Status, Habitat, etc. Resident, uncommon, subject to vertical movements. Recorded in summer (presumably breeding) between 3000 and 3600 m; descends to the foothills in winter, occasionally as low as 600 m: mixed conifer and broad-leaved forest. Habits and food as of 506 and 507. Call: a rattling *krrrr* or *chărrarŭk*; also some mellow oriole-like whistles.

509 GREENFINCH *Carduelis spinoides* Vigors   p. 224
Size Sparrow − ; length 14 cm (5½ in.).

Field Characters Male. *Above*, forehead, supercilium and sides of neck yellow forming an indistinct nuchal collar. Crown, ear-coverts, a malar streak, and back, blackish brown; rump yellow. Wings dark brown with a large yellow patch. Tail dark brown; basal half of outer rectrices yellow. *Below*, yellow. Female like male but duller and washed with green above (subspecies *spinoides*).

Status, Habitat, etc. Mostly a summer visitor. Breeds between 1800 and 3100 m; winters in the foothills mostly below 1500 and in the duars and adjacent plains. Affects open pine and deodar forest in summer, freely entering hill-station gardens. Pairs, parties and flocks. Feeds chiefly in bushes and trees, often on flower-heads. Food: seeds, berries, insects. Call: a characteristic, far-carrying *beez*; also a long-drawn whistle *weeeeeee-tu* reminiscent of iora.

510 TIBETAN SISKIN *Carduelis thibetana* (Hume)   p. 193
Size Sparrow − ; length 12 cm (5 in.).

Field Characters A small yellow-green and brownish finch with small pointed bill and slightly forked tail. Male. *Above*, olive-yellow; mantle faintly streaked with dark brown, rump brighter yellow. An ill-defined supercilium and collar bright yellow. A yellow wing-bar. Tail-feathers dark brown edged with yellow. *Below*, deep yellow, washed with olive on sides of neck and flanks. Female duller and streaked with dark brown above and below, excepting throat and upper breast.

Status, Habitat, etc. Mostly winter visitor between 1000 and 3000 m. Observed in summer (presumably breeding) in SE. Tibet between 2800 and 3800 m in hemlock and birch forest; *may* breed in adjoining northern Arunachal Pradesh. Flocks of 10 to 50. Feeds largely in tree tops; also on ground under bushes. Food: seeds. Call: a continual tremulous twittering when in flock.

511 **TIBETAN TWITE** *Acanthis flavirostris* (Linnaeus)

Size Sparrow − ; length 13 cm (5 in.).

Field Characters A small brown finch with pink rump and distinctly forked tail. Male. *Above*, fulvous-brown streaked with dark brown. Two buffish wing-bars. Rump pale pink: tail dark brown with white outer edge. *Below*, fulvous-brown: breast and flanks streaked with dark brown; belly buffish. Female similar but without pink rump (subspecies *rufostrigata*).

Status, Habitat, etc. Resident, common, subject to slight vertical movements. Breeds between 3600 and 4800 m; found in winter down to 3000 m. Affects screes and stony hillsides with *Caragana* bushes; arid boulder-strewn alpine meadows. In winter often around upland cultivation. Pairs or small flocks; large ones in winter. Runs about and feeds on ground and in bushes from flower-heads. Food: seeds. Call: a double *twite-twite* in flight.

512 **REDBROWED FINCH** *Callacanthis burtoni* (Gould)   p. 193

Size Sparrow + ; length 17 cm (6½ in.).

Field Characters Male. *Above*, forehead, supercilium, and round eye crimson. Crown black; back brown. Wings black, spotted with white. Tail black, outer rectrices and tip white. *Below*, chin and throat pinkish red; sides of throat and cheeks black. Rest fulvous-brown tinged with pinkish red. Female. *Above*, forehead, supercilium, and round eye ochre-yellow; crown dark brown; rest as in male. *Below*, ochraceous-brown.

Status, Habitat, etc. Resident? East of Nepal recorded only in Sikkim. Rare (at 3000 m in winter)—fairly open conifer forest. Small flocks of 6 to 12. Feeds mostly on ground. Food: seeds of deodar etc. Call: a loud clear whistle, not unlike bullfinch's but higher in key.

513 **PLAINCOLOURED MOUNTAIN FINCH**
    Leucosticte nemoricola (Hodgson)   p. 193

Size Sparrow; length 15 cm (6 in.).

Field Characters A high-altitude finch. Resembles a dark slim female House Sparrow, with finer bill and markedly forked tail. *Above*, brown streaked darker; a faint pale supercilium. Wings dark brown with a buff bar and buff edges to tertials. Rump grey; tail-coverts broadly white-tipped. Tail blackish. *Below*, pale grey-brown; sides of breast, and flanks, streaked with dark brown (subspecies *nemoricola*). Sexes alike.

Status, Habitat, etc. Resident, common, subject to vertical movements. Breeds between 4200 and 5300 m; winters between 4500 and 2000 m. Alpine meadows, glacier moraines and dwarf scrub above tree-line (summer); open hillslopes and fallow fields round upland villages (winter). Very gregarious. In flocks even in summer: often huge ones in winter, flying around restlessly in scattered undulating rabbles. Runs about and feeds on arid ground. Food: seeds. Call: a soft, lively sparrow-like twitter.

514 HIMALAYAN MOUNTAIN FINCH
Leucosticte brandti Bonaparte

Size Sparrow + ; length 18 cm (7 in.).

Field Characters A blackish high-elevation finch with rosy rump and forked tail. *Above*, forehead, face and crown blackish; back paler brown scalloped with sandy; rump rosy pink. A large whitish patch in wing, and pale shoulder. Tail blackish with white outer edges. *Below*, drab grey (subspecies *haematopygia* and *pallidior*). Sexes alike.

Status, Habitat, etc. Resident, common, subject to slight vertical movements. Breeds between 4300 and 5400 m; winters down to 3500 m. Affects desolate stony hillsides, moraines and alpine meadows. Gregarious at all seasons; large flocks in winter. Fond of feeding on the edge of melting snow patches and along water's edge on lake shores. Food: mostly seeds. Call: alarm, a harsh, distinctive *churr*. A loud *twitt, twitt* in flight.

515 ROSEFINCH *Carpodacus erythrinus* (Pallas)   p. 225

Size Sparrow; length 15 cm (6 in.).

Field Characters Male. *Above*, crown crimson with a paler supercilium; a dark stripe behind eye. Nape and sides of neck crimson-brown; back, wings and tail brown and crimson; rump dark crimson. *Below*, cheeks pink; chin, throat and breast crimson paling to pink on belly, whitish on vent. Female. *Above*, olive-brown with two pale wing-bars. *Below*, throat whitish with brown streaks; breast heavily brown-streaked; belly whitish (subspecies *roseatus*).

Status, Habitat, etc. Common summer (breeding) visitor. Breeds between 3000 and 4000 m; winters below 1500 m through the duars and peninsular plains. Bush-covered slopes and open coniferous forest (summer); openly wooded country and cultivation environs (winter). Pairs or flocks according to season. Food: seeds, flower buds and nectar, berries. Call: a canary-like interrogative *twee-ee?*. Song, of 5 to 8 cheery whistling notes like *twee-twee-tweeou* ..., and variations.

516 NEPAL DARK ROSEFINCH *Carpodacus nipalensis* (Hodgson)

p. 193

Size Sparrow; length 15 cm (6 in.).

Field Characters Male. *Above*, forecrown and supercilium vinaceous crimson. A wide band through eye, and upperparts, dark brown washed with crimson. *Below*, throat and belly vinaceous pink; a wide maroon breast-band; flanks dark crimson-brown. Distinguished from male *erythrinus* (515) and *rubescens* (517) by absence of red on rump. Female. *Above*, dark olive-brown streaked darker on back; two pale wing-bars. *Below*, plain olive-brown. Distinguished from all female rosefinches except *rubescens*, q.v., by unstreaked olive-brown underparts (subspecies *nipalensis*).

Status, Habitat, etc. Altitudinal migrant, fairly common: summer 3000 to 4200 m; winter mostly between 2700 and 1800 m. Rhododendron and silver fir forest, grassy slopes with stunted bushes etc. (summer); forest clearings and cultivation environs (winter). Pairs or small flocks, feeding on bushes or the ground. Food: seeds, berries, flower buds, nectar, etc. Call: a characteristic rather wailing, plaintive double whistle; a sparrow-like twitter.

517 BLANFORD'S ROSEFINCH *Carpodacus rubescens* (Blanford)

p. 225

Size Sparrow; length 15 cm (6 in.).

Field Characters Male. *Above*, crimson-brown, brighter crimson on head and rump; a double crimson wing-bar. *Below*, rosy red, grey on vent. Female. *Above*, olive-brown, slightly crimson or more olive on rump; a faint double wing-bar. *Below*, throat and breast olive-brown; belly grey; under tail-coverts edged with white. Differs from all female rosefinches (except *C. nipalensis*, 516 q.v.) in having unstreaked underparts.

Status, Habitat, etc. Resident, scarce, subject to vertical movements. Recorded in summer between 3100 and 3800 m; in winter between 1800 and 1300 m (Sikkim). Conifer or mixed conifer and birch forest. Pairs or small flocks. Feeds chiefly on ground. Food: unrecorded; presumably as of the genus. Call: a peculiar persistent clacking note.

518 **PINKBROWED ROSEFINCH** *Carpodacus rhodochrous* (Vigors)

p. 225

Size Sparrow; length 15 cm (6 in.).

Field Characters Male. *Above*, crown and a band behind eye crimson-brown; forehead and supercilium rose-pink. Back streaked with dark brown; rump rose-pink. A faint wing-bar. *Below*, entirely rose-pink. The very similar male *C. pulcherrimus* (520) has the crown streaked with dark brown. Female. *Above*, olive-brown streaked with dark brown. A prominent pale supercilium. *Below*, ochraceous buff, streaked with dark brown.

Status, Habitat, etc. Resident, subject to vertical movements. Breeds between 3000 and 4000 m; winters normally from about 3000 m down to 1200 or so. Open birch and fir forest, willow bushes and dwarf junipers (summer); open hillsides, grassy slopes and scrub jungle (winter). Pairs or flocks, feeding chiefly on ground. Food: seeds and berries. Call: a canary-like *sweet*.

519 **WHITEBROWED ROSEFINCH**

Carpodacus thura Bonaparte & Schlegel p. 193

Size Sparrow + ; length 17 cm (6½ in.).

Field Characters Male. *Above*, crown and back brown, streaked with blackish; forehead pinkish white; supercilium pink and white; lores crimson; a dark brown stripe behind eye. Rump rosy pink. Wings brown with a fine white bar and a second pink bar. *Below*, rosy pink; throat paler streaked with silky white. Glistening pink-and-white forehead, broad pink-and-white supercilium, and double wing-bar identify the male. Female. *Above*, brown streaked with blackish; a broad whitish supercilium; a fine whitish wing-bar. Rump golden yellow. *Below*, buffish, streaked with dark brown; throat rufous-buff (subspecies *thura* and *femininus*).

Status, Habitat, etc. Resident, subject to vertical movements. Breeds mostly between 3800 and 4200 m; winters from 3900 down to 2000 or so. Light fir, juniper and rhododendron forest and dwarf rhododendron around timber-line (summer); open hillsides with barberry and other scrub (winter). Pairs or loose flocks of 20 or so. Feeds on ground, walking or hopping. Food: seeds and berries. Call: a continual loud and rapid piping while feeding.

520 **BEAUTIFUL ROSEFINCH** *Carpodacus pulcherrimus* (Moore)

p. 225

Size Sparrow; length 15 cm (6 in.).

Field Characters Male. *Above*, forehead and broad supercilium rosy pink and brown. Upper plumage ashy brown streaked with blackish; rump rosy red. Two faint pinkish wing-bars. *Below*, rosy red with blackish streaks on breast and belly. Flanks brown, streaked darker. Female. *Above*, fulvous-brown broadly streaked with blackish; a faint broad fulvous supercilium. *Below*, fulvous white streaked with dark brown (subspecies *pulcherrimus* and *waltoni*).

Status, Habitat, etc. Resident, subject to vertical movements. Breeds between 3600 and 4500 m; winters from 3500 m down to 2000 or so. Rhododendron and juniper scrub around timber-line and terraced cultivation in high valleys (summer); open scrub-covered hillsides (winter). Small parties, feeding on the ground. Habits etc. as of the genus. Food: seeds and vegetable matter. Call: a sparrow-like *cheet-cheet*.

521 **LARGE ROSEFINCH** *Carpodacus edwardsii* Verreaux   p. 193
Size Sparrow + ; length 17 cm (6½ in.).
Field Characters Male. *Above*, forehead mixed pink and crimson-brown; a broad glistening pink supercilium. Crown and nape deep crimson-brown with black shaft-lines. Rest blackish and rufous brown washed with crimson; rump unstreaked rufous brown with faint crimson flush. A double pinkish wing-bar. *Below*, cheeks, chin and throat pink with black shaft-streaks; rest rosy brown, bright rose-pink on abdomen and vent. Female. *Above*, ashy ochre-brown with broad dark streaks; a faint buffish supercilium. *Below*, buff, finely streaked with brown (subspecies *rubicunda*).

Status, Habitat, etc. Resident, uncommon, subject to vertical movements. Breeds between 3400 and 4000 m; winters from 3700 m down to 2000, rarely lower. Rhododendron and silver fir forest (summer); open forest and mountainsides with rose, barberry and other scrub (winter). Pairs or small parties. Feeds on ground or in low bushes. Food: seeds and berries. Call: undescribed.

522 **REDBREASTED ROSEFINCH** *Carpodacus puniceus* (Blyth)
p. 225

Size Bulbul ± ; length 20 cm (8 in.).
Field Characters An extreme high-elevation finch. Male. *Above*, forehead and supercilium bright scarlet. Crown and back grey-brown streaked with blackish; a broad brown band behind eye. Rump rosy red. Wings and tail brown. *Below*, throat and breast scarlet mottled with white; rest grey-brown. Female. *Above*, grey-brown streaked with blackish; rump olive-yellow. *Below*, cream-buff boldly streaked with blackish (subspecies *puniceus*).

Status, Habitat, etc. Resident, common, subject to vertical movements. Breeds between 4200 and 5200 m; observed in winter between 4700 and 2700 m. Frequents steep rocky and boulder-strewn slopes in the alpine zone. Small parties feed on ground on hillsides, commonly amongst melting snow. Food: chiefly seeds; also flower buds and petals. Call: a metallic sparrow-like chirp in flight; a cheery bulbul-like *Are you quite ready?*

523 CROSSBILL *Loxia curvirostra* Linnaeus p. 225
Size Sparrow; length 15 cm (6 in.).

Field Characters Easily identified by unique structure of bill in which the tips of the mandibles cross each other. Male. *Above*, crown and back orange-red mottled with blackish; a blackish band through eye; rump bright orange. *Below*, orange-red. Female. *Above*, streaked dark brown with olive-yellow wash; rump yellow. *Below*, greyish, washed with olive-yellow, especially on breast (subspecies *himalayensis*).

Status, Habitat, etc. Erratic: imperfectly known. Recorded between 2700 and 4000 m at all seasons; in winter exceptionally down to 1500 m (Sikkim). Conifer forest. Small restless flocks. Keeps to tops of larch, pine and fir trees, clinging to the cones in acrobatic positions to extract the seeds with the specially adapted bill. Food: seeds of conifers. Call: series of 3 to 5 *kip-kip-kip-kip* while feeding and in flight.

524 REDHEADED ROSEFINCH
Propyrrhula subhimachala (Hodgson) p. 225
Size Bulbul; length 20 cm (8 in.).

Field Characters Male. *Above*, forehead and supercilium crimson; crown, nape, sides of neck, and back brown washed with crimson. Shoulder and wing-bar reddish. Rump crimson. *Below*, chin, throat and breast crimson with whitish spots; rest greyish brown. Female. *Above*, forehead and supercilium orange-yellow; crown and back scalloped brown and olive. Nape and sides of neck olive. Rump yellowish olive. *Below*, throat and breast yellow with dark mottling; rest grey.

Status, Habitat, etc. Resident, uncommon, subject to vertical movements. In summer between 3500 and 4200 m; in winter from 3500 down to 1800 or so: thick rhododendron, juniper and willow scrub near timber-line; also light forest with dense undergrowth. Small flocks in winter, feeding in bushes. Food: pine seeds, berries, flower buds and petals. Call: a sparrow-like but melodious chirruping.

525 SCARLET FINCH *Haematospiza sipahi* (Hodgson) p. 193
Size Bulbul ± ; length 18 cm (7 in.).

Field Characters Male. Brilliant scarlet overall, with brown wings and tail. A blackish line through eye. Bill yellow. Female. *Above*, brownish olive-yellow; rump bright yellow (conspicuous in flight). *Below*, brown scalloped with yellowish grey.

Status, Habitat, etc. Resident, uncommon, subject to vertical movements. Breeding zone imperfectly known: recorded in summer at 2300–2400 m, in winter down in the foothills and duars—open pine and oak forest and tropical jungle. Scattered flocks. Feeds in bushes as well as high trees. Has strong and dipping, typical finch-like, flight with rapid wing beats. Food: seeds, berries, flower buds, insects. Call: a pleasant soft high-pitched *too-ee*.

526 GOLDHEADED BLACK FINCH

Pyrrhoplectes epauletta (Hodgson)   p. 193

Size Sparrow; length 15 cm (6 in.).

Field Characters Male. Entire plumage black except hindcrown and nape, which are golden orange. Middle of belly orange-buff. A distinctive white line on wing. Female. Head and neck yellowish ashy grey. Rest of plumage chestnut-brown, paler on underparts. White line in wing as in male.

Status, Habitat, etc. Resident, fairly common, subject to vertical movements. Occurs in summer between 2900 and 3900 m; in winter from 3600 down to 1400 m. Rhododendron forest with ringal bamboo undergrowth (summer); low scrub, dense thickets and bushes in forest (winter). Small parties in winter, often associated with rosefinches. Feeds in bushes or on ground. Food: seeds and berries. Call: a distinctive, high-pitched squeaky whistle *peeu peeu*.

527 BROWN BULLFINCH *Pyrrhula nipalensis* Hodgson   p. 225

Size Sparrow + ; length 17 cm (6½ in.).

Field Characters Male. *Above*, a dark brown band around base of short, abrupt, swollen bill. Crown scaly ashy brown; back, wing-coverts and sides of neck brown. A white spot under eye. Rump purplish black with a narrow white band posteriorly. Wings and tail glossy purplish black with velvety black tips to the feathers. A whitish patch on wing. A thin crimson streak on innermost tertials. *Below*, pale brown, whitish on belly. Female similar but streak on tertials yellow instead of crimson (subspecies *nipalensis*).

Status, Habitat, etc. Resident, fairly common locally: subject to vertical movements. In summer between 2100 and 3000 m, exceptionally to 4000 (Sikkim); in winter down to 1500 m or so. Dense fir, oak, and rhododendron forest. Entirely arboreal: forages high up in trees. Pairs

or small flocks. Food: seeds, berries, flower buds, nectar. Call: a
mellow *pearl-lee*; a continual soft whistling while feeding.

528 **BEAVAN'S BULLFINCH** *Pyrrhula erythaca* Blyth   p. 225
Size Sparrow + ; length 17 cm (6½ in.).
Field Characters Male. Distinguished from Redheaded Bullfinch (529)
by ashy grey crown like back (*v.* orange-chestnut). Upper rump black,
lower rump white. *Below*, throat ashy grey; breast, belly and flanks
orange-red. Lower belly grey paling to white on vent. Female browner
and duller without red on breast. White forecrown distinguishes her
from female *P. nipalensis* (subspecies *erythaca*).
Status, Habitat, etc. Resident, locally common, subject to vertical
movements. In summer between 2500 and 3800 m; in winter from at
least 3200 m down to 2000 or slightly lower. Conifer and rhododendron
forest; also willow and buckthorn thickets. Habits etc. as of the genus.
Pairs or small flocks, feeding among bushes or on ground. Food: seeds,
berries, flower buds, nectar. Call: a triple whistle; rather similar to call
of Redheaded (529).

529 **REDHEADED BULLFINCH** *Pyrrhula erythrocephala* Vigors

p. 193

Size Sparrow + ; length 17 cm (6½ in.).
Field Characters Male. *Above*, forehead, and around base of bill and eye,
velvety black; crown, nape and hindneck orange-chestnut; back ashy
grey; rump white. Tail (noticeably forked) and wings (with a large ashy
patch and wing-bar) glossy purplish black. *Below*, chin black; throat,
breast, and flanks dull rusty red fading to ashy on belly, white on vent.
Female. Crown yellowish olive; back and underparts brown.
Status, Habitat, etc. Resident, fairly common, subject to vertical
movements. Breeds between 2700 and 3900 m: some descending in
winter as low as 1000 m. Birch, willow and rhododendron forest
(summer), rhododendron, oaks and conifers (winter). Habits and food
as of other bullfinches. Call: a soft plaintive whistle *phew-phew* similar
to European bullfinch's.

BUNTINGS: Emberizidae

530 **CHESTNUT BUNTING** *Emberiza rutila* Pallas   p. 225
Size Sparrow ± ; length 14 cm (5½ in.).
Field Characters A slim sparrow-like bird with noticeably forked tail.
Male. *Above*, entirely chestnut. *Below*, throat chestnut; rest sulphur

yellow. **Female**. *Above*, crown ashy brown with dark streaks. A broad brown band on sides of crown; a buffish streak behind eye; dark ear-coverts. Back ashy brown with blackish streaks; rump chestnut. *Below*, throat buffish, with dark malar stripes; breast yellowish olive finely streaked. Rest yellow, streaked blackish on flanks.

Status, Habitat, etc. Scarce winter visitor from SE. Siberia; recorded from Jalpaiguri duars and Sikkim, in rice stubbles, forest clearings and bushes in cultivation. Flocks in winter. Feeds on ground, flies up into trees when disturbed. **Food**: chiefly grass seeds. **Call**: a thin high *teseep* when flushed.

531 YELLOWBREASTED BUNTING *Emberiza aureola* Pallas
Size Sparrow; length 15 cm (6 in.).

Field Characters **Male** (winter). *Above*, crown, nape, sides of neck, back, and rump chestnut, the feathers pale-fringed. A large white shoulder-patch and narrow white wing-bar. Tail blackish; outer rectrices largely white. *Below*, yellow, streaked with blackish on flanks. A prominent chestnut collar on upper breast. **Female**. *Above*, brown streaked with blackish; some chestnut on rump. A buff supercilium and dark ear-coverts. White in wing-pattern replaced by buff. Tail and underparts as in male but without the chestnut breast-band (subspecies *aureola*).

Status, Habitat, etc. Winter visitor from E. Siberia. Fairly common in the more easterly Arunachal plains, duars and up to 1500 m: grassland, and cultivation around villages. Habits, food, etc. as of other buntings. **Call**: a short *zipp* and a soft trilling *trssit*.

532 BLACKFACED BUNTING *Emberiza spodocephala* Pallas
Size Sparrow; length 15 cm (6 in.).

Field Characters **Male**. *Above*, base of bill, lores, and around eye black; head and nape olive-grey; back brown with blackish streaks; two buffish wing-bars. White outer tail-feathers conspicuous in flight. *Below*, chin black; throat and breast olive-grey; belly pale yellow. **Female**. *Above*, crown rufous-brown streaked with dark brown like back; a pale supercilium and dark ear-coverts. *Below*, pale yellow with dark moustachial streaks; breast and flanks streaked with blackish (subspecies *sordida*).

Status, Habitat, etc. Winter visitor to the Jalpaiguri and Sikkim duars: high grass and scrub jungle, hedgerows around cultivation, damp paddy stubbles, etc. Small loose flocks feeding on ground. **Food**: paddy grains and grass seeds. **Call**: normal note *tsik*.

533 ROCK BUNTING *Emberiza cia* Linnaeus
Size Sparrow; length 15 cm (6 in.).

Field Characters M a l e. *Above*, head bluish grey with deep chestnut coronal stripes; a black stripe through eye; black moustachical streak looping up to join eye-stripe behind cheeks; cheeks and supercilium white. Back chestnut-brown streaked with black; rump rufous-chestnut. Outer tail-feathers white, conspicuous in flight. *Below*, throat and breast bluish ashy; rest rufous-chestnut. F e m a l e similar but duller (subspecies *khamensis*).

Status, Habitat, etc. Common resident, in northern Arunachal, presumably occurring in winter in Sikkim and Bhutan; subject to vertical movements. Breeds between 2700 and 4200 m, winters from 3300 down to 2200 m or so: dry rocky, bush-covered mountainsides. Solos, pairs, or small flocks. Feeds on ground, perches on bushes. Tail constantly flicked open and shut, flashing the white outer rectrices. F o o d: seeds, grain, insects. C a l l: an oft-repeated subdued squeak *tsi* or *swip*. A twittering goldfinch-like song.

534 GREYHEADED BUNTING *Emberiza fucata* Pallas
Size Sparrow; length 15 cm (6 in.).

Field Characters M a l e. *Above*, crown and nape grey, streaked with black; ear-coverts chestnut. Back rufous-brown, heavily streaked with black; rump rufous. Wings brown; tertials black, edged with rufous. Wing shoulders rufous; outer rectrices white. *Below*, throat and upper breast white with dense black streaking forming a breast-band; black moustachial stripes. Lower breast chestnut; flanks rufous-chestnut streaked with blackish; belly whitish. F e m a l e similar but paler and duller. Crown pale brown (*v.* grey); a short whitish supercilium (subspecies *fucata*).

Status, Habitat, etc. Winter visitor from Amurland, Manchuria, Japan. Locally common: duars and foothills—in wet stubbles, marshy grassland and bushes. Habits, food, etc. as of other buntings. C a l l: not specifically recorded in winter.

535 LITTLE BUNTING *Emberiza pusilla* Pallas   p. 225
Size Sparrow − ; length 14 cm (5½ in.).

Field Characters M a l e. *Above*, crown black with a rich rufous median band or 'centre parting'; a pale rufous supercilium. A black line behind eye looping round rich rufous ear-coverts. Rest of upperparts rufous brown streaked with blackish. Tail blackish, noticeably forked, with white outer rectrices. *Below*, chin, cheeks and sides of throat pale rufous; rest white, streaked with black on breast and flanks. F e m a l e similar but duller.

Status, Habitat, etc. Winter visitor from northern Eurasia; common.

Plains, duars and foothills, mostly below 1800 m, in open country—rice stubbles, and grass and scrub around cultivation. Singles or small flocks, commonly with tree pipits and other buntings. Feeds on ground close to cover, hurriedly diving behind bushes on alarm. F o o d : seeds and insects. C a l l : a short *tsit*.

536 **CRESTED BUNTING** *Melophus lathami* (Gray)   p. 225
Size Sparrow; length 15 cm (6 in.).
Field Characters M a l e . Entirely glistening black with chestnut wings and tail, and a long pointed black crest. F e m a l e . *Above*, olive-brown with dark brown streaks; crest shorter; a pale eye-ring. Edge of wing and outer rectrices rufous. *Below*, yellowish buff with dark streaks on breast and dark moustachial stripes.
Status, Habitat, etc. Resident, locally common but capricious. In summer up to 1800 m, withdrawing mostly to lower levels and duars in winter. Stony, sparsely scrubbed hillsides and dry-deciduous jungle; partial to charred patches after fires. Small loose flocks. Habits, food, etc. as of other buntings. C a l l : a repeated *pink* while feeding. Song, a spirited whistling *which ... which ... which-whi-whee-which* (accent on *whee*), reminiscent of song of Indian Robin (*Saxicoloides fulicata*).

INDEX

Numbers in bold type refer to pages facing illustrations

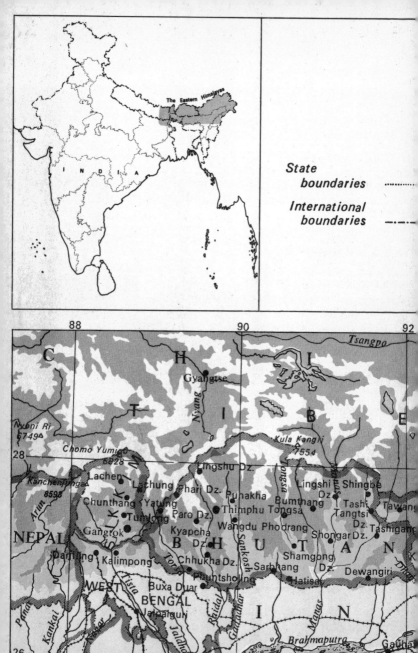

State
boundaries ....

International
boundaries —·—·—